The World Aluminum Industry in a Changing Energy Era

Publications from

The Mineral Economics and Policy Program of
Resources for the Future and
The Colorado School of Mines

General Editors

Hans H. Landsberg
Resources for the Future

John E. Tilton
The Colorado School of Mines

Material Substitution: Lessons from Tin-Using Industries
John E. Tilton (1983)

State Mineral Enterprises: An Investigation into Their Impact on International Mineral Markets
Marian Radetzki (1985)

Metallic Mineral Exploration: An Economic Analysis
Roderick G. Eggert (1987)

The Mining Law: A Study in Perpetual Motion
John D. Leshy (1987)

The World Aluminum Industry in a Changing Energy Era
Merton J. Peck, ed. (1988)

World Mineral Exploration: Trends and Economic Issues
John E. Tilton, Roderick G. Eggert,
and Hans H. Landsberg, eds. (1988)

The World Aluminum Industry in a Changing Energy Era

Merton J. Peck, editor

A Project of
Resources for the Future and
The Colorado School of Mines
in cooperation with
The International Institute for
Applied Systems Analysis

Resources for the Future ▪ Washington, D.C.

Printed in the United States of America

Published by Resources for the Future
1616 P Street, N.W., Washington, D.C. 20036
Books from Resources for the Future are distributed worldwide by The Johns Hopkins University Press.

Library of Congress Cataloging-in-Publication Data

The world aluminum industry in a changing energy era / Merton J. Peck, editor; Resources for the Future and the Colorado School of Mines in cooperation with the International Institute for Applied Systems Analysis.
p. cm.
Includes bibliographies and index.
ISBN 0-915707-42-X (alk. paper)
1. Aluminum industry and trade—Case studies. 2. Aluminum industry and trade—Energy consumption—Costs—Case studies. 3. Aluminum industry and trade—Government policy—Case studies. I. Peck, Merton J. II. Resources for the Future. III. Colorado School of Mines. IV. International Institute for Applied Systems Analysis.
HD9539.A62W67 1988 88-4990
338.2'74926—dc19 CIP

∞ The paper in this book meets the guidelines for permanence and durability of the Committee on Production Guidelines for Book Longevity of the Council on Library Resources.

RESOURCES
FOR THE FUTURE

RESOURCES FOR THE FUTURE (RFF) is an independent nonprofit organization that advances research and public education in the development, conservation, and use of natural resources and in the quality of the environment. Established in 1952 with the cooperation of the Ford Foundation, it is supported by an endowment and by grants from foundations, government agencies, and corporations. Grants are accepted on the condition that RFF is solely responsible for the conduct of its research and the dissemination of its work to the public. The organization does not perform proprietary research.

RFF research is primarily social scientific, especially economic. It is concerned with the relationship of people to the natural environmental resources of land, water, and air; with the products and services derived from these basic resources; and with the effects of production and consumption on environmental quality and on human health and well-being. Grouped into five units—the Energy and Materials Division, the Quality of the Environment Division, the Renewable Resources Division, the National Center for Food and Agricultural Policy, and the Center for Risk Management—staff members pursue a wide variety of interests, including forest economics, natural gas policy, multiple use of public lands, mineral economics, air and water pollution, energy and national security, hazardous wastes, the economics of outer space, and climate resources. Resident staff members conduct most of the organization's work; a few others carry out research elsewhere under grants from RFF.

Resources for the Future takes responsibility for the selection of subjects for study and for the appointment of fellows, as well as for their freedom of inquiry. The views of RFF staff members and the interpretations and conclusions of RFF publications should not be attributed to Resources for the Future, its directors, or its officers. As an organization, RFF does not take positions on laws, policies, or events, nor does it lobby.

This book is a product of the Energy and Materials Division of Resources for the Future, Joel Darmstadter, director. Merton J. Peck is Thomas DeWitt Cuyler Professor of Economics at Yale University and acting dean of the Yale School of Organization and Management.

The book was edited by Leah Mazade and designed by Joan Engelhardt. The index was prepared by Florence Robinson.

CONTENTS

FIGURES

TABLES

FOREWORD

With this volume, Merton Peck and his collaborators sustain a long-time record of involvement by Resources for the Future in research on economic issues in world aluminum. There is, of course, nothing mysterious about the organization's concern with this commodity: the fact of an electricity-intensive production process defines aluminum's significant link to energy resources, while, in its multiple fields of application—including those in which it competes with other materials—aluminum has been an important force for technological improvement in numerous product lines. Both aspects comport with RFF's traditional interest in illuminating issues of resource adequacy and management.

Prominent among the present volume's forebears is Sterling Brubaker's *Trends in the World Aluminum Industry* (published for RFF by Johns Hopkins Press in 1967). In many ways the stirring perceived by Brubaker two decades ago, on the part of developing economies especially, to move from mining ore to producing finished aluminum metal has progressed in a very significant way. The factors responsible for that development are among the topics fully articulated by the contributors to this volume. At a more discrete level of analysis, the matter of aluminum's market share in one particular—albeit important—product line was among the topics addressed in RFF's 1983 publication *Material Substitution: Lessons from Tin-Using Industries,* edited by John Tilton.

The World Aluminum Industry in a Changing Energy Era succeeds, I believe, in once again advancing understanding of this major player on the world resources scene. RFF, through its Mineral Economics and Policy Program, is pleased to have facilitated the underlying research for this effort and to be the publisher of the resulting study.

Joel Darmstadter
Director, Energy and Materials Division
Resources for the Future

PREFACE

This book was made possible by the energy and initiative of John Tilton, now professor of mineral economics at the Colorado School of Mines. As head of the Mineral Economics programs at the International Institute for Applied Systems Analysis, he planned the project and has been its shepherd since its inception in 1983. His advice and encouragement have been extremely helpful to all the authors.

The International Institute for Applied Systems Analysis and Resources for the Future provided financial support. The Institute also sponsored two conferences in Vienna for the participants in the project. Joel Darmstadter and Hans H. Landsberg of the staff of Resources for the Future provided detailed criticisms, as did the three reviewers.

The editor was assisted by Yale students Christopher Erickson, Mark Hulak, and Amon Liu. Virginia Casey and Judith Imperati typed successive drafts. The editor and authors are extremely grateful for all the willing and expert help that made this book possible.

Merton J. Peck
Yale University

February 1988

1

INTRODUCTION

MERTON J. PECK

Readers of this volume know that the energy shocks of 1973 and 1979 triggered significant changes throughout the world economy. What is less widely recognized is that the oil shocks, by affecting the price of electricity generation, dramatically altered the international competitiveness of industries whose production processes use large amounts of electricity. Furthermore, the energy shocks did not increase the price of electricity in all countries equally because some nations are blessed with sufficient hydropower or low-cost coal to prevent electricity prices from rising as sharply as in nations more dependent on oil-generated power. As a heavy user of electricity the primary aluminum smelting industry is a leading example of the effects of such variations in energy costs.

By 1985 the energy situation had changed dramatically with oil prices falling by a third from their 1980 peak. There were predictions that prices over the next five years would be stable or would decline even further.[1] Research for this book was carried out in 1982 when the decline in oil prices had only begun. Still, the price changes of 1984 and 1985 do not mean a return to the cheap energy of the 1970s: in real terms the price declines still leave 1985 oil prices about 70 percent higher than their 1972 level. For the energy-intensive aluminum smelting process the energy price increases

Merton J. Peck is acting dean of the Yale School of Organization and Management and Thomas DeWitt Cuyler Professor of Economics at Yale University.

Editor's note: In this book, metric tons (tonnes) are used unless otherwise indicated.

that have occurred since 1972 have been significant; more importantly, the differences in electricity prices among regions remain. Hence, the restructuring of the primary aluminum industry—that is, the shift of production from one region to another—that began in the 1970s is likely to continue.

Restructuring has occurred mainly in the six nations that account for about 90 percent of primary aluminum production. (Primary aluminum is produced from bauxite ore, in contrast to secondary aluminum, which is produced from scrap.) The simple story told in this volume is that with the rise in energy costs, three regions—Japan, the United States, and Western Europe—have become high-cost locations for primary aluminum production relative to three others—Australia, Brazil, and Canada (the so-called ABC countries). These changes are largely the result of the post-1972 differences in the price of electricity. But there is more to the story than that, and it involves the complications and qualifications we shall describe in the next six chapters.

Many of these complications arise from the impact of public policy on the aluminum industry. Accordingly, the chapters describe and present an analysis of the public policy choices made in each country. The policy focus is appropriate because electric power—the key input for aluminum smelting—is produced either by government-owned or government-regulated suppliers. In most countries, aluminum smelters are one of the biggest industrial users of electric power. In making decisions about electricity rates, governments are inevitably making decisions about the aluminum industry: whether to promote its growth in the low-cost power countries and whether to maintain its existing size in the high-cost power countries. Governments also influence their aluminum industries by the more general measures of taxes, exchange rates, tariffs, and industry subsidies. The provision of electric power, however, is the element that makes government involvement in the aluminum industry distinctive.

Our primary objective in this volume is to provide information and analysis to those interested in the world aluminum industry—whether they be business executives, policymakers, or economists. But we also have a broader goal: to provide a case study of the process of structural adjustment in industrialized economies.

All national economies must continually adjust to new conditions. Sometimes those conditions are shifts in consumer demand, sometimes new technology, and sometimes changes in international competitiveness. The message of various Organisation for

Economic Co-operation and Development (OECD) position papers on positive adjustment is clear: there is a great deal at stake and much to be gained from carrying out structural adjustment relatively quickly and painlessly. Without a fairly smooth working process of structural adjustment, the growth that has been occurring in world living standards would slow down.[2] The international economy that has evolved since World War II would be threatened if trade barriers arose to protect high-cost industries and domestic production were substituted for international specialization and trade.

Indeed, structural adjustment, broadly defined, is one of the two major economic issues of the 1980s. The other is macroeconomic stabilization—that is, achieving full employment and price stability. To policymakers, the macro problems always seem more immediate and dramatic; but a longer view would give priority to the problems of an economy in promoting the industries in which it is internationally competitive and in phasing out those in which it has lost its competitiveness.

Our study examines both ends of the process of adjustment—that is, both winners and losers—and consequently is a useful case study of structural adjustment. It also illustrates the difficulties of structural adjustment for countries that have gained and those that have lost their international competitiveness. Furthermore, this study of six nations permits a comparative analysis of public policy regarding the aluminum industry. Although we give no formal marks to the various nations, we examine the costs and benefits of their policies and the political and economic forces that underlie them. A study of all six of the nations that make up most of the world aluminum industry underlines the proposition that, in the long run, individual national policies are interdependent: what one nation does sets limits on what other nations can do. If the nations that lose international competitiveness maintain or even expand their aluminum industries, those that have gained competitiveness will lack the markets to expand their industries. Among the nations that have gained competitive advantage, there is intense rivalry, for if one nation expands at a rapid rate, less of a market would remain for the others. It is generally assumed that the world aluminum market will coordinate and harmonize the decisions of various national aluminum industries. But with public policy playing such a major role, the proposition that the market will bring about an efficient allocation of aluminum production among nations must be verified empirically rather than being assumed.

The six chapters in this volume vary in style, focus, and scope—an expected outcome with academic economists as authors. Yet they each describe the postwar evolution of their national aluminum industry, its international competitiveness, and public policy toward it. The rest of this introduction gives an overview of the industry, in terms of its structure and the costs of smelting; examines the overall character of the structural adjustment process; and summarizes the six chapters, particularly with respect to public policy.

THE STRUCTURE OF THE INDUSTRY AND THE COSTS OF ALUMINUM SMELTING

The Stages of Aluminum Production

The authors of this volume examine aluminum smelting, the stage of production that uses the most electricity. The chapters also refer to other production stages, however, and readers unfamiliar with the industry may find a brief description of these stages useful.

Figure 1-1 shows the main features of the industry; the square boxes indicate processes, and the rounded boxes indicate products. Starting at the top the diagram shows that bauxite is mined from ore deposits and converted first into alumina by the Bayer process. Alumina is then converted into primary aluminum using the Hall-Heroult process. This is the smelting stage that is the subject of our study.

Through a variety of processes common to metalworking—rolling, casting, extruding, and so forth—primary aluminum is used to make fabricated aluminum products. Fabrications are used to manufacture items ranging from window frames, household foil, and engine blocks to aircraft wings. In terms of use, aluminum is second only to steel.

Like most metals industries the aluminum industry generates scrap in its production operations. The scrap is sent to secondary smelters that remelt it to make secondary aluminum, which can be substituted for primary aluminum in most uses. Another source of scrap is discarded final products (for example, beverage cans), which are collected and sorted by dealers who sell the scrap to secondary smelters.

Primary and secondary aluminum are close substitutes, and together they form a pool of metal for fabricating. Secondary aluminum's share of the total pool has been growing in recent years, however. For example, from 1970 to 1980 its share of U.S. domestic primary and secondary production rose from 20 to 25 percent; by

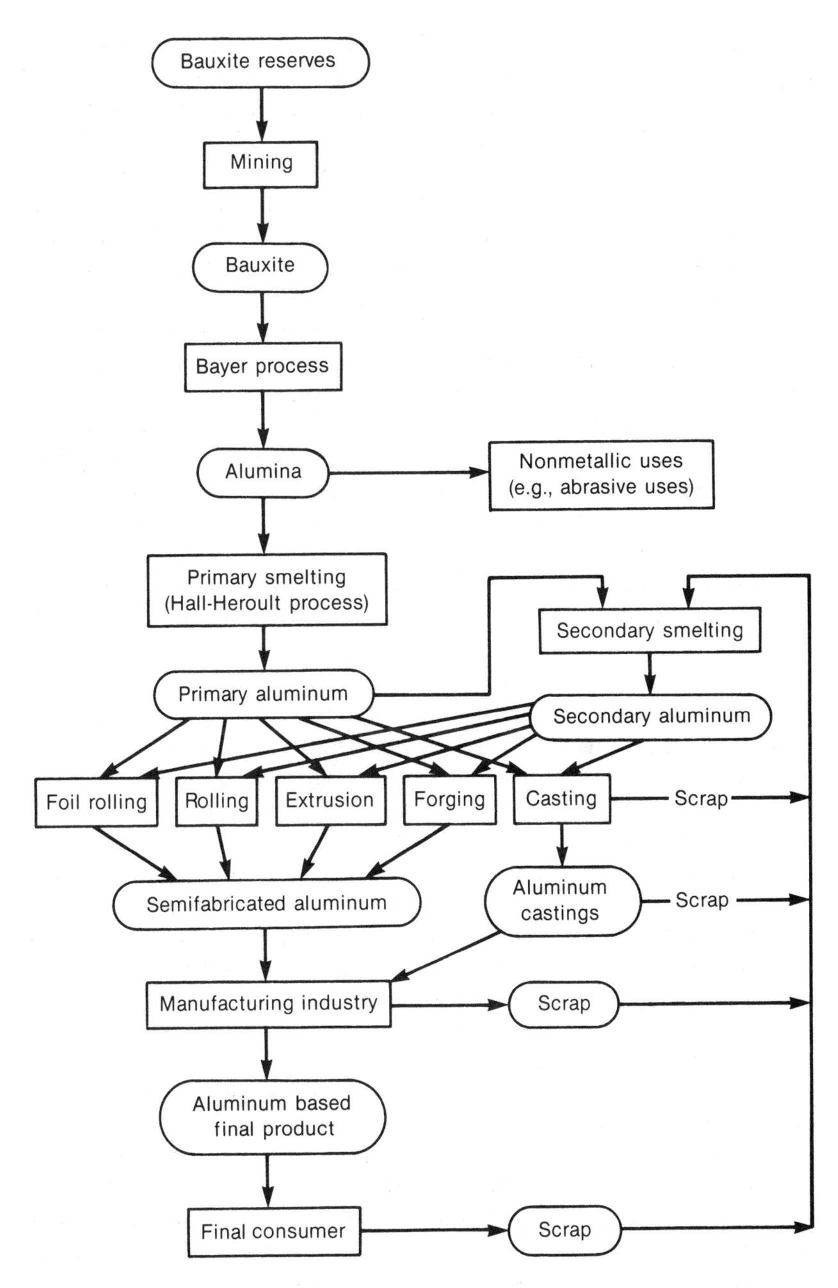

Figure 1-1. Principal stages and flows in the production of aluminum products. *Source:* Based on Organisation for Economic Co-operation and Development, *Aluminum Industry: Energy Aspects of Structural Change* (Paris, OECD, 1984) p. 93, with additions by the author.

1984 it had risen to 30 percent.[3] The relative importance of secondary production is determined partially by the rate of growth of aluminum consumption. Secondary output is set in part by the primary production of a decade or two earlier; a high rate of growth in the total demand for aluminum reduces the proportion of secondary aluminum because there are limits on the recycling of scrap from discarded end products. (New scrap generated in the fabricating process grows proportionally with fabricating production.) The volume of secondary aluminum is also influenced by the price of primary aluminum because higher prices for primary aluminum lead to more recovery of scrap. Finally, secondary recovery is sensitive to the uses to which aluminum is put, beverage cans having the shortest life and aluminum siding one of the longest. All three of these factors account for the growing importance of secondary aluminum since 1980, although primary aluminum clearly remains the dominant product.

One striking feature of the aluminum industry is that the Bayer process for alumina developed in 1888 and the Hall-Heroult smelting process for aluminum invented in 1886 have been in use since the founding of the industry. Technological change has occurred since then and plays a role in the various country studies in this volume, but it has taken the form of successive improvements in the two existing processes as opposed to the development of entirely new processes. Of particular concern to this study is the progressive reduction in electricity requirements for aluminum smelting that has resulted from these improvements. In the United States, for example, energy requirements per pound of aluminum produced declined by about 1 percent annually from 1947 to 1972; they declined 0.5 percent annually from 1972 to 1980. The slowdown in the improvement of energy consumption from 1972 to 1980 can be attributed to low rates of new smelter construction; many of the technological improvements can be economically incorporated only in newly constructed smelters.[4]

The two energy shocks renewed interest in developing radical alternatives to the energy-intensive Hall-Heroult process. These efforts are concentrated on obtaining aluminum directly from bauxite, thus bypassing the electrolytic reduction stage. Such efforts have a long history dating back to pilot plant operations in World War II Germany. After the war, the Aluminum Company of Canada (Alcan) and Aluminium Pechiney of France carried on experiments but were unable to develop a successful alternative. In the 1970s the Aluminum Company of America (Alcoa) established an operating smelter using chlorine reduction, but Alcoa closed down the project after fifteen years of experimentation. Several companies have experimented with partial carbothermatic reduction of silicon-rich aluminum ores in an arc furnace. According to recent assessments, however, "the extreme complexity of the reactions involved, and the severity of conditions encountered, require extensive development which will take many years to accomplish."[5]

Technological developments are notoriously difficult to predict. Past records suggest that continued research on the existing Hall-Heroult process will reduce energy requirements, particularly for new smelters. Technical barriers, however, may continue to preclude radical new technologies that could significantly alter the energy intensity of aluminum reduction.

The various stages of aluminum production are sufficiently separate and distinct that they often occur in different countries. Over half of the bauxite used to make alumina comes from developing countries, principally Caribbean, Latin American, and African nations. Only two of the countries in this study, Australia and Brazil, produce enough bauxite to satisfy their needs; the others rely largely on imports. At the alumina stage, production begins to shift from bauxite-producing countries to industrialized nations, which have about half the total alumina refining capacity; the other half is in the bauxite-producing countries. At the smelting stage, the major consuming countries dominate with about 70 percent of primary aluminum production located in the United States, Canada, Europe, and Japan. The fabrication of aluminum generally takes place close to its consumption. Three regions—the United States, Japan, and Western Europe—produce and consume most of the world's fabricating output.

Although the various production stages are technologically and geographically distinct, they are often linked economically by vertical integration (that is, the common ownership of various stages of production). Products are also transferred from one stage to another through long-term contracts, a method that is often termed quasi-vertical integration. In an excellent study of vertical integration, Stuckey estimated that in the late 1970s, 80 percent of the bauxite mined for use in the industry moved to alumina refineries as intracorporate transfers, 10 percent moved through long-term contracts, and only 10 percent moved as spot sales. Of the alumina subsequently sent to smelters, 85 percent went as intracorporate transfers, with the remaining 15 percent divided between sales under long-term contracts and spot sales.[6] More recent percentages are not available, but there appears to be relatively little change. Vertical integration is still important in the link between the smelting stage and fabrications, but at this point in the production process market transactions predominate, though there is considerable variation among countries. Only in the final product stage does vertical integration become unimportant.

The Structure of the Industry

As of 1960 six firms accounted for most of the world's primary aluminum production (although there were twenty other primary producers):[7] Alcan, Alcoa, Kaiser Aluminum and Chemical Company (Kaiser, headquartered in the United States), Pechiney, Reynolds Metals Company (Reynolds, in the United States), and Schweizerische Aluminium AG (Alusuisse, in Switzerland). Since the 1960s, however, the "big six's" share of primary aluminum production has steadily declined; by 1981 it was only 53 percent of the primary aluminum capacity in the market economies (that is, exluding the USSR, Eastern Europe, and China)[8] This decline reflected the entry of new producers: the number of aluminum producers worldwide increased from 26 in 1960 to 71 in 1981.[9] Many of the entrants were state owned; as a result, by 1981 partly or wholly state-owned enterprises accounted for 46 percent of world aluminum capacity. (This percentage includes Pechiney, which was nationalized in 1981.)

Despite the entry of more firms the big six remain significant, and the reader will encounter these companies in all but one of the six chapters that follow. The three big-six companies headquartered in the United States account for 60 percent of U.S. capacity.[10] Alcan has recently acquired a U.S. smelter; Pechiney and Alusuisse have recently sold their small interests in U.S. smelting capacity. In Western Europe, all six own smelters, with the extent of their ownership varying considerably by country. In Australia, the big six are all active and account for most of the smelting capacity, usually through joint ownership with local interests. In Brazil, both Alcoa and Alcan have smelters. In Canada, Alcan was dominant until 1980, owning all but one of the smelters. Since 1980, however, the remaining members of the big six have all been involved in plans for building Canadian smelters. Japan is the sole exception to big-six activity; all of its primary producers are Japanese-based companies.

The existence of the big six may have a special significance for structural adjustment to changes in international competitiveness.

Because they operate in many countries, these companies can locate new smelters in areas that have the best prospects for the lowest cost operations with an ease that a strictly national company cannot match. They can also concentrate production at those of their existing smelters with the lowest operating costs, irrespective of national location. Their ability to concentrate production in this way is limited, however, by start-up and shutdown costs, union contracts, and power contracts with minimum consumption provisions

Because of the international scope of their operations, the big six are also positioned for effective bargaining on favorable power contracts and other concessions when decisions on the location of new smelters are made. They can choose fairly freely among countries, although competition for the scarce supply of low-cost power limits their bargaining power. It is unclear, however, just how much big-six status increases such power. As the following chapters report, primary producers other than the big six have begun to build smelters outside their national locations, giving these newcomers worldwide options as well. The Japanese aluminum producers are the most striking example of such a trend with a very large smelting project in Brazil. The issue of bargaining for smelter locations is discussed subsequently in this introduction, as well as later in the volume.

The Costs of the Smelting Process

International competitiveness in the long run is primarily determined by the costs of production. It is now fashionable to say that international competitiveness or comparative advantage does not exist: it is created by company and governmental actions. Such a statement is only partially true for aluminum smelting. Comparative advantage in smelting is related to the availability of low-cost power, which depends in turn on the endowments given to a region by nature relative to the demand for electric power from users. Still, it falls within the realm of public policy to decide whether to develop low-cost power sources and, once developed, whether to use the power for aluminum smelting. And it is a public policy decision to raise the price existing smelters must pay for electricity and so reduce their international competitiveness. We devote much attention in the chapters that follow to the impact of public policy for electricity prices on the costs of aluminum smelting.

Aluminum smelting is a capital-intensive production process. Consequently, there is a significant difference between two definitions of costs: unit operating costs and total unit costs. Unit operating costs are the costs that can be avoided by not operating the smelter. (They are sometimes called redfield costs or variable costs.[11]) In order of general importance, the principal components of aluminum smelter operating costs are alumina, electricity, labor, and other raw materials.[12] Total unit costs include capital costs, as well as items like property taxes and insurance, which apply whether the smelter operates or not.

Operating costs are the costs primary producers consider when deciding whether or not to operate a smelter. A simple statement of a producer's decision criterion is that a smelter is operated if its unit operating costs are equal to or less than the current price for aluminum. Unless that condition is met the firm can reduce its losses by closing the smelter. It is often profitable for a firm to operate a smelter even though the price of aluminum is less than the *total* unit cost (the smelter in that instance operates at a loss by conventional accounting standards). This situation arises because, as noted above, the total unit cost includes costs that are incurred whether the smelter operates or not. Hence, the loss would be even greater if the smelter were not operated. Because aluminum smelting is a capital-intensive process, the gap between unit operating and total unit costs is substantial. As a broad generalization, about a third of the total unit costs are for capital, insurance, and property taxes that continue whether the smelter is operated or not. These sunk costs are thus irrelevant to an operating decision.

The decision criterion for smelter operation is stated here in an oversimplified form because certain operating costs are partially fixed in the short run. For example, electric power and alumina are often bought under contracts that require minimum purchases, and union contracts often impose cost penalties for layoffs. Furthermore, there are costs, sometimes as much as several million dollars, in closing and opening a smelter. As a result, a smelter will continue to operate, even if the price of aluminum is below its operating costs, if the firm regards the current price as only a short-term condition.

The operating costs of existing smelters vary considerably, primarily because of variations in the price of electricity. In 1984 the worldwide range in smelter power costs was 4.5 to 55 mills (1 mill = 0.1 cent), which translates into a range of 3.5 to 43 cents of electric power cost per pound of aluminum.[13] Two other items also vary by smelter. Power requirements and worker-hours per pound of aluminum can vary by about 30 percent, with older smelters generally having higher requirements. Wage rates also differ among countries, but this difference has relatively little impact on cost variations.[14]

The operating or variable costs of the 122 smelters in the world can be arranged from lowest to highest to produce the supply curve shown in figure 1-2. (Note that figure 1-2 is a supply curve for the industry and not the more familiar cost curve for a single plant. Hence, there are no economies of scale.) The supply curve is of considerable importance because the spot price of primary aluminum fluctuates considerably; in 1984, for example, the price varied from a low of 42 cents to a high of 97 cents (in current U.S. dollars).[15] At the low price, hardly any of the smelters covered their operating costs; at the high price, almost all did. Smelters with very high unit operating costs are called swing capacity smelters and are usually operated only when prices are high. There are exceptions, however, as discussed in the United States and Western European chapters.

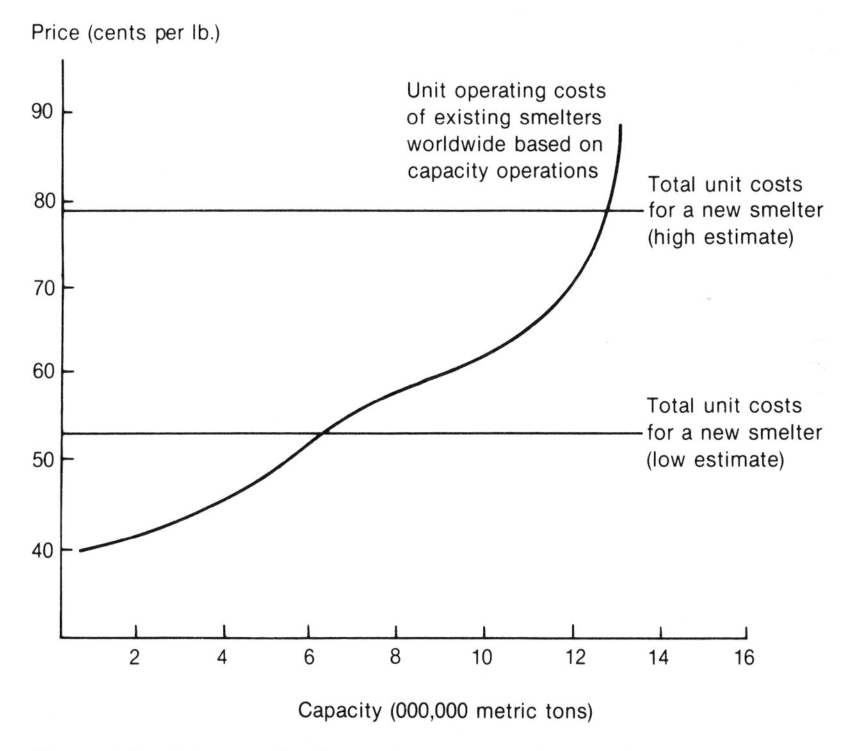

Figure 1-2. Primary aluminum cost curves. *Source:* Operating costs are from U.S. Department of Commerce, "Energy and the Primary Aluminum Industry" (mimeo, Washington, D.C., 1984) p. 53, based on data from Anthony Bird Associates. Total unit costs for a new smelter are based on table 7-3, footnote (a), of this volume. A tonne (metric ton) equals 1.1 English tons.

Furthermore, figure 1-2 applies to the early 1980s; since then, primary aluminum producers have permanently closed some smelters, and they have made major efforts to reduce costs at their remaining facilities. The first action would eliminate the upper tail of the distribution; the second would shift the cost curve downward.

A different set of considerations applies to the decision to build a new smelter. A firm will build a smelter if the expected revenues over the life of the smelter will cover total costs, including a profit on the invested capital. As shown in figure 1-2, that decision criterion can be translated into an expected price that will cover total unit costs. The supply curve for new plants is a horizontal line because, at any given point in time, a primary producer will seek out the current low-cost locations to build an aluminum smelter that applies the latest technology and that is built at a scale that will result in the lowest costs. The wide range shown in figure 1-2 reflects variations in the published estimates of the costs for new smelters.

New capacity may be constructed to meet an increased demand for aluminum or to replace existing high-cost capacity. The decision criterion for building a replacement smelter is that the total unit costs for the new smelter must be lower than the operating costs of the smelter it replaces. The wide range of estimates shown in figure 1-2 suggests considerable uncertainty as to whether it would be profitable to build new capacity to replace existing smelters. At the low estimate, half the existing smelters would be replaced; at the high estimate, only a few would be replaced.

In an important respect, however, both of the figure's estimates of total unit costs are misleading: they are estimates of the cost of building only one new smelter. Because the number of low-cost locations at any one time is limited, the total unit cost curve rises with a construction program that involves building many smelters at the same time, some of which would have to be built at higher-cost locations. Because firms tend to avoid building smelters at such locations, the availability of low-cost sites constrains the rate at which new smelters could replace existing ones.

The actual behavior of the primary producers indicates how the pattern of investment decisions changes over time. When prices (and hence demand) are high, smelters are built to serve the additional demand; when prices and demand are low, smelters with high operating costs are closed—first temporarily, and then, if the downturn is long-lived, permanently. The end result is a gradual displacement of high-cost smelters.

The situation is further complicated by the long planning and construction period for a new smelter and the fact that low-cost locations are not always available. As they become available, largely because of government power projects, primary producers tend to build new smelters regardless of other considerations because if they do not, the opportunity may pass or be taken up by another producer. The availability of new sites with low-cost power is one key determinant of new smelter construction, although if a downturn develops once the smelter is started, the project is often delayed.

The differences between the criteria for operating existing smelters and building new ones result in two standards for international competitiveness: one for the operation of existing smelters and a second for the construction of new smelters. A nation can lose its international competitiveness for the building of new capacity and still remain internationally competitive for the operation of existing smelters. Furthermore, the variation in the operating costs of existing smelters within a nation means that some of its smelters may be internationally competitive while others are not.

Although the most significant distinction in costs is between those of existing smelters (redfield or variable costs) and the construction of new smelters (greenfield costs), there is an intermediate case: the expansion of existing smelters (brownfield costs). Historically, this has been an important way of adding capacity in the aluminum industry.

Brownfield costs are similar to greenfield costs in that total unit costs (including those of capital) are relevant. Expanding an existing smelter is often cheaper than constructing a new one—that is, up to some point usually set by such constraints as the supply of low-cost power, the capacity of the transportation system, the availability of land and labor, and, recently, potential environmental problems.[16] Smelters are often planned at the outset for expansion; the typical practice is to build a smelter with 60,000 to 100,000 tons of capacity and then to expand it until capacity reaches 200,000 tons or greater.[17] Initially, a company generally limits its investment because larger increments of capacity would represent a major expansion of an individual firm that might result in idle capacity. (Even for Alcoa, the largest producer during the 1950s, a 100,000-ton addition would have been a 10 percent increase in the company's capacity at that time.) Adding capacity in successive stages allows a firm to expand more nearly in line with its sales.

Economies of scale in aluminum smelting for up to 100,000 tons of capacity are reported in numerous engineering studies.[18] By 1972, after considerable brownfield expansion and a slowdown in building new plants, only four of the thirty-two existing U.S. smelters had less than 100,000 tons of annual capacity. Seventeen of the smelters had a capacity of over 150,000 tons; the largest had a capacity of 305,000 tons.[19] The actual pattern of capacity suggests that economies of scale may well exist beyond the 100,000 tons mentioned repeatedly in the literature on the aluminum industry.

Electric Power

As indicated earlier, the price of electric power is the major determinant of the international competitiveness of smelters. Electricity is usually the second largest item in terms of cost (alumina ranks first), but its importance for competitiveness derives from the fact that it is the cost that varies most by geographic location.

Electric power is produced in several ways: oil thermal generation, coal thermal generation, nuclear power, or hydropower. Each power source has a different level of costs, ranging roughly from 65 mills per kilowatt hour for oil to as low as 15 mills for hydropower.[20] Of course, hydropower can be produced only at specific locations with large flows of water; in contrast, oil and nuclear power generating plants can be located almost anywhere. (Coal, with its high transportation costs, is an intermediate case.) Although electricity increasingly can be transported considerable distances, it is still consumed most economically in the region in which it is produced. The economic consequence of these technological facts is that the lowest cost smelters are located largely in regions with hydropower.

There are other complications that affect the price of electricity, however. For example, there are very large economies of scale in the construction of power facilities, economies that are most marked with hydropower and nuclear power plants and least with oil generation but that are significant in all forms of generation. This means that, when demand grows, regions often have power shortages at the going prices until new power capacity is added. The additions often result in more power than can be used, creating a power surplus. Because the operating costs of hydropower facilities are low—about 10 percent of total costs—there is a strong incentive with a power surplus to attract smelters at rates that will make some contribution to the fixed capital costs, even though those rates are usually well below those charged other customers and below the total (including capital) cost of generating electricity.

The growth of demand in a region may convert a power surplus into a shortage at some future time, making low rates to smelters an economic burden. If the power supplier then raises its rates to the smelter, the smelter is converted from a low-cost operation into a high-cost one, and it may no longer be internationally competitive. A further complication is that hydropower capacity varies with the flow of water, which can change from year to year. In addition, the demand of other consumers for power may change with variations in the level of economic activity in the region. Thus, surpluses and shortages are difficult to predict: the demand–supply balance for electric power can change markedly from year to year, and even the average in future years is unpredictable.

Analyzing the determinants of the price of electric power can be approached in two quite different ways: (1) as an examination of the bargaining between the power supplier and the aluminum producer to determine power rates to smelters or (2) as the application of the criteria of welfare economics to determine the optimum rates for overall economic efficiency. The first approach is positive economics—a characterization of what actually happens; the second is normative economics—what electricity rates to smelters should be. Both approaches are used in the chapters that follow, and they are described in detail there. Here, we discuss only the general features of each.

Electric Power Bargaining

The kind of bargaining that occurs between a primary aluminum producer and a power supplier depends on whether the power will supply a new or an existing smelter. With a new smelter the primary producer can compare the power costs at alternative locations and bargain with each potential supplier of power—even though they may be thousands of miles apart. For a nation to attract a smelter, it must offer power rates competitive with alternative locations. It is not surprising, then, that the ABC countries (Australia, Brazil, and Canada) are attracting new smelters; the authors of this volume suggest that rates for new smelters in those countries tend toward a common level of around 14 to 16 mills per kilowatt hour. But the rates cited in the chapters are only approximations. Because the power suppliers are often governments, the bargaining often includes provisions for taxation, local financing, and government-furnished infrastructure; high electricity prices can thus be offset by liberal provisions in other areas. Even the actual prices paid for electricity are often confidential. The OECD study cited earlier points out "the lack of transparency of power prices and the pricing mechanism governing power supply contracts to smelters."[21] (There is also considerable variation by nation in the public availability of power rates paid by smelters. In the United States, most power contracts are on file with regulatory agencies.)

Even though each party to the power bargain has alternatives, those alternatives are limited on both sides of the market. At any one time, there are only a few primary producers seeking to expand their smelting capacity and only a few suppliers who can offer large blocks of power. Small numbers of buyers and sellers lead to a range of indeterminancy in the terms of sales; as a result, an important factor is the particular circumstances of each sale, such as estimates of the parties as to the future demand for power from other consumers or the projections of the demand for aluminum.

Both parties are interested in electricity rates over the life of the smelter. But the primary producer is at a disadvantage in that once the smelter is built, the producer loses a great deal of its bargaining power because it can no longer consider options elsewhere. Primary producers have used three methods to offset their loss of bargaining power once they have built a smelter:

1. They can own the power facilities. This option requires large capital outlays, particularly for hydropower plants. Furthermore, building dams requires government permission, which is sometimes not easily forthcoming.
2. They can sign long-term contracts that guarantee low-cost power. This is the most common method for capitalizing on the producers' strongest period of bargaining power—when they are choosing among greenfield sites. Yet, such contracts do not provide complete certainty because they can be abrogated by governmental action.
3. They can diversify the location of their smelting among regions, among suppliers of power, among power sources, and finally among nations. With such a diversity of sources, actions related to a particular power supply will affect only some of the smelters of a primary producer and so place the producer at less of a disadvantage vis-à-vis his competitors. But diversification sometimes means giving up the lowest cost green-

field opportunity. Diversification is also, of course, an option most open to the big six.

Each of these options has disadvantages; none of them provides certain protection. The risk that power prices will change despite long-term contracts has become particularly great in the late 1970s and 1980s. Before that time, power costs for aluminum smelters were usually set by long-term contracts. But the decade of the 1970s saw rapidly rising electricity prices, reflecting the increases in all energy costs following the two oil shocks. The already substantial gap between the electricity prices paid by smelters under long-term contracts and those paid by other consumers became even larger, and the rates to smelters were often raised despite the contracts.

Primary producers also bargain with power suppliers over the rates paid by existing smelters. Their bargaining power in this instance comes from their ability to close a smelter if electricity rates make its operations uncompetitive. Because the smelter is usually a large consumer of power, closing it can convert a region from one of power shortage to one of surplus. In addition, the closure will lead to the unemployment of the smelter workers, a prospect that may tip the political process in favor of electricity rates that permit continued operation of a smelter. A primary producer may also choose the intermediate course of making a smelter swing capacity, that is, operating it only when the demand for aluminum is high.

The political influences in this whole bargaining process are also quite important because power suppliers are often government agencies or, at the least, regulated private companies. Thus, the suppliers may have explicit objectives beyond selling power, objectives such as promoting regional development or maintaining employment. And they may be particularly sensitive to price differences among consumers, particularly differences between prices for industrial consumers and those for household consumers—who, after all, are the most numerous voters.

Later chapters in this volume describe such bargaining over electric power. Perhaps the only generalization possible is that the outcomes are diverse and that the bargaining often determines the international competitiveness of a nation's aluminum industry.

Normative Standards for Electric Power Rates

Although some part of each of the following chapters is concerned with the realities of bargaining over power, the preacher in every economist leads to a discussion of normative standards. The fundamental question is whether the actual and potential shifts in the location of aluminum smelting correspond to the criteria of economic efficiency. Given the importance of electricity prices, the bargaining that determines them, and political concerns about maintaining particular national industries, this is clearly a significant question. As the OECD study states, "The main question [of structural adjustment] is whether the energy market signals reach the [aluminum] industry without being too much distorted by power pricing policies applied to smelters."[22]

But the idea of distortion implies a norm. The norm applied here is that of welfare economics in which prices should maximize economic efficiency, which can be defined as an allocation of resources among various productive activities that best serves consumers. Welfare economics holds that prices should equal the marginal social opportunity cost of each product. Such costs are determined by the prices paid to capital and labor to attract them from alternative uses. Thus, a product's marginal cost reflects what consumers give up in the form of alternative uses of the factors of production. The willingness of consumers to pay the incremental cost of additional production shows that they value its production over that of alternative products.

The marginal social opportunity cost as applied to electricity prices for smelters has two different meanings: (1) the value of electricity to other users, as reflected in the prices they are willing to pay for an incremental unit of electricity, and (2) the cost of producing an incremental unit of electricity. The first definition applies when increasing the supply of electricity is so costly that it is not a realistic option; the second definition applies when an increase in the supply of electricity is feasible.

The second concept of marginal social opportunity cost for a supplier of electric power is illustrated in figure 1-3. The cost curve is shown in two segments: a low-cost segment representing a low-cost source of power (usually hydropower) and a high-cost source (usually thermal power). The low-cost source has a finite capacity; once it has been fully used, further supplies of electric power must come from the high-cost source.

The marginal social cost of our illustrative utility depends on the location of the demand curve. If the demand curve intersects the cost curve in the region of the low-cost supply, D_1D_1, then P_1 is the price that equals the social cost. But if the demand for power increases so that the demand curve shifts to D_2D_2, then P_2 represents the marginal social cost. This situation is not just a theoretical

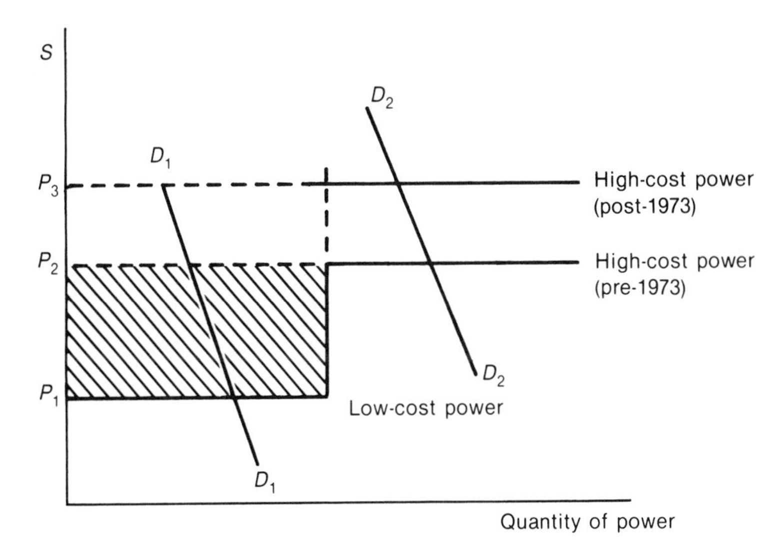

Figure 1-3. Alternative electric power situations.

curiosity; it is rather common, as can be seen from the discussions in the chapters. As economic growth occurs, it shifts the demand curve in the figure to the right (that is, toward increased demand) as both households and industries consume more power. This entire situation was intensified by the oil shocks, which raised the cost of thermal power; as shown in figure 1-3, P_3 is the price that equals marginal social cost after 1973. The point of this illustration is simply that marginal cost pricing creates a large surplus or rents from the assistance of low-cost power as demand for the high-cost power source increases. These surpluses arise in the power sector, however, and not in the aluminum industry itself. Economists thus would classify the surplus as rents to owners of scarce factors—that is, low-cost power—whose supply cannot be increased.

Elsewhere in a market economy, rents go to whoever happens to be the owner of the scarce resource, whether the item in question is a Rembrandt or land in midtown Manhattan. Electric power, however, is a government-owned or government-regulated enterprise, and as a result the concept of ownership is not straightforward. Although sometimes the surplus is retained by private investors or the taxpayers, more often it is used to reduce electricity prices. Sometimes priority is given to reductions for residential users; at other times it may be given to industrial users; and at still other times it may be used to establish a price between P_1 and P_2 for all consumers. Another practice is to charge P_1 to established customers (which would include existing aluminum smelters) and P_2 to new customers. Finally, certain classes of consumers may be singled out for low rates—for example, charging P_1 for reasons of economic development or maintaining employment or, alternatively, to give low rates to households.

The aluminum industry's large consumption of electricity makes the industry a major factor in decisions on power rates. Its competitiveness is largely determined by the policy on electricity prices adopted by its suppliers. Strict adherence to marginal cost pricing would make many smelters uneconomical; yet departures from that criterion open the door to extensive subsidization.

The situation is even more complex than figure 1-3 indicates. The figure assumes that the shift of the demand curve is known and occurs once and for all. But as the following chapters indicate, the demand curve for power can shift (to the left in the figure), toward decreasing demand in response to recessions, creating power surpluses at previously established prices. Power facilities have a long gestation period, and investment decisions are based on demand forecasts. Many forecasts made in the 1970s have turned out to be overly optimistic, leaving the suppliers of power with a substantial surplus. The question then becomes who should receive rate reductions to stimulate power consumption, which in turn becomes intertwined with the uncertainty of estimates of the duration of the power surplus.

The careful reader might look now for a discussion of other major cost items, particularly bauxite and its derivative alumina. But the search will be in vain, for this reason: even though alumina is usually the largest cost item, it is easily transportable and so does not vary markedly by geographic location. Hence, the story of shifting international competitiveness can be largely told with reference only to national differences in electricity prices. The dominance of electricity prices is demonstrated in Carmine Nappi's essay on Canada (chapter 7), a country relying totally on imported bauxite. The essay on Brazil by Eliezer Braz-Pereira (chapter 6) describes the opposite case, a bauxite-rich country in which the development of aluminum smelting still depends on low-cost hydropower. There are no cases in which cheap bauxite offsets high-cost power.

Aluminum Prices

The decision criteria we have just enumerated turn on the relationship of prices to costs. The behavior of prices for primary aluminum changed radically in the 1970s from a pattern of stable, so-called administered prices to one of commodity prices set in commodity exchanges similar to those established for wheat in the early 1980s. Prior to 1980, list prices, those quoted by primary producers, moved serenely upward each year; spot prices, those available from metal traders and brokers, varied around the list prices. The primary producers did not alter their published ingot prices with each variation in the spot price but instead made unpublished concessions that made their actual transaction price close to the current spot price. In general, the gap between list and spot prices was small. After 1980, however, the picture changed. The gap between list and spot prices became substantial, and almost all sales were made at spot prices. The spot price itself was very volatile, as figure 1-4

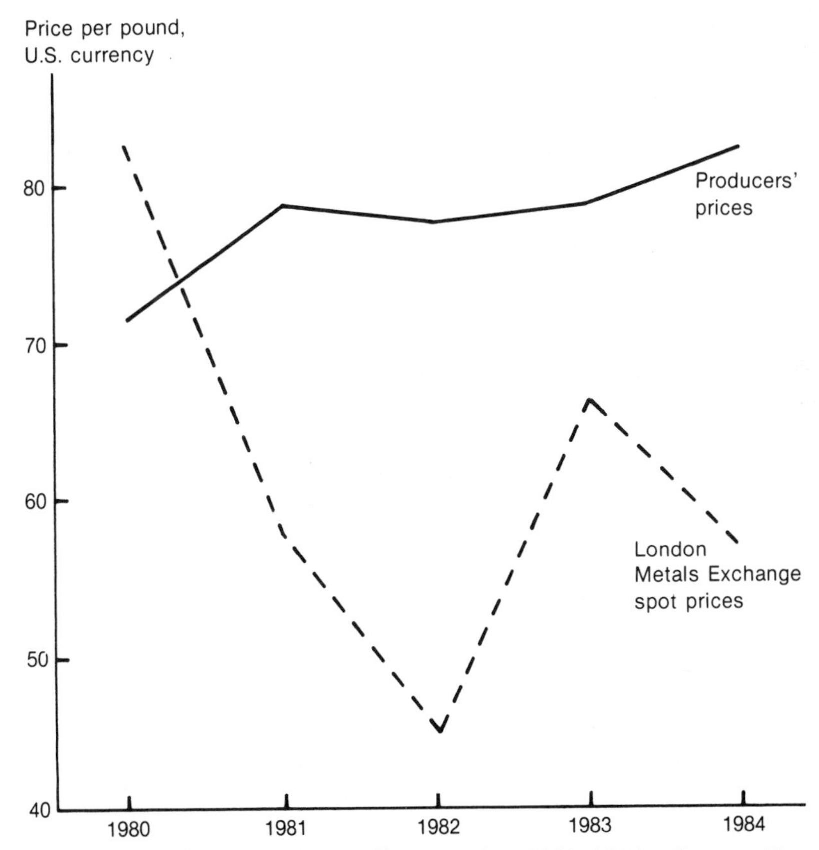

Figure 1-4. Aluminum price fluctuations, 1980–1984. *Source:* Producers' prices are from *Metal Statistics 1985: The Purchasing Guide of the Metal Industries* (New York, Fairchild Publications, 1985) p. 20. Spot prices were calculated from the same source, p. 24.

shows. (The U.S. spot price follows the London Metals Exchange price.)

The shift from administered to commodity prices reflected the decline, noted earlier, in the dominance of the industry by the big six producers, with the immediate cause being the slowdown that occurred in the growth in aluminum demand. The transition to a regime of flexible prices appears to be permanent. In a move that symbolized the shift, American producers discontinued publishing list prices in July 1984.

This shift in aluminum pricing has occurred in an industry in which many of the costs—such as electric power—are set by long-term contracts. In such an environment, primary producers can easily be caught in a price squeeze; to prevent that outcome, as the essays report, producers are beginning to negotiate electric power contracts that tie the price of electricity paid by smelters to the commodity price of aluminum. That approach, however, violates the welfare norm that electricity prices should be based on marginal social costs of electricity production. Instead, the utility now shares the risks of price fluctuations in the aluminum market. The violation of the welfare norm does not mean that the current pricing of power is wrong; the welfare norm may simply be a standard that is too simple for a complex world and therefore unworkable today.

THE ADJUSTMENT PROCESS: AN OVERALL PERSPECTIVE

Prior to 1980, the construction of new capacity to serve a growing demand was the main method by which the geographical distribution of aluminum capacity changed. That has not been the case since 1980, however; demand has been in constant flux, but the trend in aluminum consumption has been flat, showing practically no growth. The hope remains that the no-growth era reflects only the recessions of the early 1980s and that in due course aluminum consumption will resume its growth. Forecasts of the long-run growth of aluminum consumption, however, are not clear about such a trend and in fact show wide variation. The OECD assembled the forecasts of six leading organizations; these organizations project annual growth rates of 3 to 5 percent for the latter part of the century. The percentage difference is small, but the magic of compound interest is such that even a fraction of a percentage point makes a sizable difference. A 3 percent growth rate would require construction of new capacity in 1993; a 5 percent growth rate would require new capacity in 1990.[23]

In this volume, we are not concerned with the hazardous business of forecasting demand. Nevertheless, the growth of demand will determine in part the rate at which new capacity is constructed in the countries with available low-cost power. The process is not a mechanical one but responds to the differences in electricity prices among nations. Table 1-1 shows how substantial those differences have become. (The reader should note that the table illustrates only general magnitudes and that in high-cost locations there is often little long-term incremental power available.) Major concessions on power costs thus can speed up the process of building new smelters.

The essay on Canada (chapter 7) demonstrates how responsive new capacity construction can be to changes in electricity prices. Hydro-Quebec, Canada's leading hydropower producer, offered 8-mill power to new smelters for the first five years of their operation, leading to a wave of planning for new smelters. Brazil made comparable concessions, as chapter 6 recounts. Because the particular arrangements in the two countries are sometimes tailor made for each company and involve long-term contracts, exact comparisons are not possible. Still, there is obviously a market for new capacity, albeit an imperfect one, and concessions in one country tend to be matched in others. Such a competitive process can accelerate the relocation of smelting capacity.

The other way in which smelting capacity shifts its geographic distribution is by the closing of smelters in high-cost countries.

Table 1-1. Price of Electricity to Aluminum Smelters

Locations	Mills per kilowatt-hour
Low-cost power locations	
Australia	15–19
Brazil	13–17
Canada	8–16
High-cost power locations	
United States	25–38
Western Europe:	
France	26
F. R. Germany	25
Italy	35
Norway	20
Greece	15
Japan	56

Source: Data are from OECD, *Aluminum Industry: Energy Aspects of Structural Change* (Paris, OECD, 1984) pp. 33–44, as modified by data in individual country essays. Prices are as of 1982 and are in 1982 dollars.

Japan is the best example with more than half of its 1975 capacity closed by 1983. The high price of electricity reported in table 1-1 makes that outcome no surprise. As chapter 4 reports in detail, however, a government–industry plan for the closing of the smelters provided government aid to share the losses. Concurrently, Japanese producers began building smelters in Brazil; as a result, the shift from high-cost to low-cost locations was partly internalized within the individual firms.

In the United States the highest cost smelters have been closed, but there are others that are on the margin of international competitiveness. Their fate depends on future decisions on electricity rates, which are complicated in the United States by the great diversity of suppliers of electric power. Some suppliers have given high priority to keeping their smelter customers in operation by granting concessions on electricity rates; others have not. Christian Kirchner's essay on Western Europe (chapter 3) also stresses the diversity of the situations of various smelters, but in this case the diversity stems largely from national differences in the supply of electric power. The region as a whole has lost international competitiveness, but few smelters have closed. In contrast to Japan, Western European smelters have received subsidies or concessions on electricity prices that have permitted their continued operation.

PUBLIC POLICY AND THE ALUMINUM INDUSTRY

The world aluminum industry was particularly vulnerable to the energy shocks of the 1970s, not only because of its extensive use of energy but also because of the great expansion of the industry in the 1960s in countries that even then were high-cost locations. The expansion reflected a view that it was important to locate smelters close to the market—that is, the fabricators in the industrialized countries. The lower level of energy prices and their smaller variation among locations in the 1960s meant that the cost penalty of a location close to the market was small. In addition, at that time there were large infrastructural costs to locating production in Brazil and Australia, two of the low-cost power countries; in Canada, the third of the ABC countries, Alcan dominated the industry. All of these factors may have discouraged earlier shifts to low-cost locations.

Public policy also played a role in concentrating investment in the United States, in Western Europe, and particularly in Japan. Akira

Goto (chapter 4) points out that the growth and continued existence of the Japanese primary aluminum production capacity required tariff protection, an undervalued yen, and low-priced oil. The rounds of tariff reductions and the rise in the yen in the early 1970s made the industry noncompetitive even before the oil shocks. In Western Europe the promise of low-cost nuclear power led to various policies to encourage the expansion of smelting capacity, particularly in the Federal Republic of Germany and the United Kingdom. In the late 1970s, when nuclear power turned out to be much more expensive than anyone had forecast, the aluminum smelting industry might well have been noncompetitive even without the oil shocks. In the United States, smelter capacity expansion occurred mainly because more firms were entering the market in response to the high growth rate of aluminum demand. This concentration of investment was perhaps the least affected by public policy.

In Japan and Western Europe, public policy after the oil shocks had to cope with the question of how to deal with an industry that had become internationally noncompetitive. As indicated earlier, the Japanese response was to close smelters under an explicit policy that involved measures to shift the resulting losses to the economy generally rather than have them borne entirely by aluminum producers and their workers. Losses were shared among the shareholders of the aluminum companies, the industrial groups to which these companies belonged, the financial institutions that had made loans to the companies, and the government. Incentives or pressures to reduce capacity were provided both by market forces and subsidy schemes. Although the reductions in capacity may not have proceeded as fast or been as extensive as the loss in international competitiveness required, the adjustments were surely in the right direction.

Goto describes the various factors that brought about this policy, which can be classified as adjustment assistance rather than subsidies to maintain the industry. It may well be that relying on market forces alone would have brought about an even quicker adjustment by the industry to the energy shocks. In that case, however, the concentration of losses among the aluminum companies alone might have resulted in an effort to obtain subsidies to keep the smelters in operation.

The case of Western Europe represents just such a policy—the use of subsidies to offset and nullify the loss in international competitiveness and continue the operation of existing smelters. The rationale for such a policy is primarily to maintain employment, but in some countries, particularly West Germany, there are additional justifications; smelting operations may provide demand for high-cost coal or nuclear power generation. Apparently, aluminum fabricators also favor domestic sources of supply (which is not surprising when the costs of such a supply are subsidized).

Subsidizing an industry that has lost its international competitiveness is an inefficient use of resources. Although there are social costs from unemployment that could justify the subsidization of the smelters, that argument could be applied to almost any industry. It is peculiarly puzzling to find it applied to aluminum smelting, which is not very labor intensive.

The difference in policy response between Japan and Western Europe probably is a reflection of general economic policy. Japanese industrial policy is more explicitly aligned with the promotion of resource reallocation to favor industries that are internationally competitive. In contrast, Western European policy is more concerned with maintaining employment in specific industries and in using subsidies to achieve that objective. There are, of course, wide variations among nations and industries in Western Europe. And as the current disputes over agricultural, coal, and steel subsidies in Western Europe show, there are indications of change. But the case of the aluminum industry suggests that Japan has a comparative advantage in the flexibility with which it alters its economy to fit its changing international competitiveness.

Much U.S. policy is the result of decentralized decisions by the various suppliers of electric power to aluminum smelters, decision in which the suppliers have used various pricing policies although the price of power has climbed sharply, resulting in the closing of some smelters. That situation may be changing, however, as electricity prices have continued to increase. More smelters are closing, and the suppliers of power are considering special rates to keep their smelter customers in operation. In contrast to Japan, there has been no federal policy in this country to assist the aluminum industry in its adjustment.

An observer might think that formulating public policy in the ABC countries would be simpler because it is easier to adjust to winning than to losing. Each of these countries apparently had large supplies of low-cost power and thus was in a position to attract smelting capacity. And yet in each of the three, the aluminum industry is a subject of considerable controversy.

The textbook theory of international economics assumes that factor markets—markets for the resource industry uses, labor, capital, land, and energy—are competitive. But the key factor market in aluminum smelting is the one for electric power, which is highly imperfect on both sides of the market. On the seller's side are the power agencies in the three low-cost power countries; on the buyer's side are the few large multinational aluminum companies. There are rents to be gained on the limited number of low-cost power sites. The combination of rents, imperfect competition, and public agencies is bound to lead to controversy.

The competition among countries for smelters can drive the price of electric power below its long-run marginal opportunity cost. Nappi suggests that this outcome may have occurred in Canada, while Braz-Pereira points out that some Brazilian economists think this may have occurred for the very large-scale power projects in Northern Brazil. A related issue is whether countries that have gained competitive advantage have given away too much to the industry. Have they, in Nappi's phrase, "left money on the table"? That question has arisen in Canada because Hydro-Quebec has been eager to use cheap hydropower as an instrument of economic development. The policy question Nappi examines is whether in the long run this policy serves the best interests of the province.

For Brazil, the investment in smelters is part of a major development effort for the northern part of the country. The project involves the construction of bauxite mines, alumina plants, and smelters as well as transportation facilities and even housing, and it extends to iron mining and other industries. Aluminum projects, however, are only a small part of the strategy for the development of northern Brazil, the "Great Carajas Program," involving an investment of $62 billion.

Given Brazil's general economic difficulties in recent years, it is not surprising that some of the projects have been delayed. And as Braz-Pereira reports, there has been considerable criticism of the export-oriented, capital-intensive character of the development plan. The aluminum projects have not been spared that criticism. Even so, they have moved toward completion, perhaps because the close link to the world aluminum and capital market provided by the participation of Japanese and big six companies has insulated the aluminum projects from the country's more general economic problems.

Australia raises a quite different set of questions. The country is generally classified as one that has gained international competitiveness. Certainly the pattern of new aluminum smelter construction suggests that this is so. Yet the essay by John Beggs (chapter 5) questions this common view. Beggs states: "It is possible that Australia did not at any time have a significant comparative advantage in aluminum smelting. It is also possible that today electricity tariffs are artificially low as a result of gross errors in the cost allocation procedures that led electricity authorities to set tariffs that do not reflect the true costs of capital being employed in electricity generation."[24]

It may also be that Australia had a competitive advantage but managed to bargain it away. As Beggs points out, "A shift in comparative advantage always creates economic rents and a scramble among market participants to capture those rents."[25] In Brazil and Canada, the process of dividing the rents has been limited to the supplier of power and the primary producers. The unique feature in the process in Australia was the addition of labor unions as a significant rent-seeking participant—so significant, in fact, that Beggs concludes that Australia's competitive advantage may well have been squandered.

Perhaps the only general conclusion about public policy toward structural adjustment in the aluminum industry is that policy is shaped by national institutions and the direction of general economic policy, as well as by the particular position of the national aluminum industry in the world market. This conclusion forms a basis for urging the reader to proceed to the six country essays.

But perhaps a bit more discussion is warranted. The study of the aluminum industry demonstrates the importance of factor markets and government intervention in those markets. Economic purists may bemoan such intervention, but it may actually have a less distorting impact on trade than those older instruments of trade policy, tariffs and quotas. The resource misallocations described in the chapters seem modest when compared, for example, with those occurring in agriculture, an area in which traditional protectionism operates. And it is interesting to note that despite the great changes in international competitiveness, the commitment of nations to free trade has been strong enough that no quotas for primary aluminum have been established nor have tariffs been raised.

The ordering of the six chapters deserves mention. The United States, which is first, is a case study of a country that has lost competitiveness for the construction of new smelters and may be in the process of losing competitiveness for the operation of existing smelters. Public policy is mixed and highly decentralized, but its general direction seems to be toward accepting the loss of the country's primary aluminum industry. Western Europe follows because the economic situation of its aluminum industry is similar to that of the United States; in this case, however, public policy has attempted to keep existing smelters operating. Japan is the extreme case among these three, phasing out most of its smelters.

The next three chapters are the ABC countries: Australia, Brazil, and Canada, in that order. All have been locations for new smelters in the 1970s. The ordering of the three is not merely alphabetical; they are in reverse order of their apparent international competitiveness. Placing Canada last allows the volume to end on an optimistic note.

NOTES

1. John C. Sawhill and Richard Cotton, "Introduction," in John C. Sawhill and Richard C. Cotton, eds., *Energy Conservation: Successes and Failures* (Washington, D.C., The Brookings Institution, 1986) p. 2.
2. Organisation for Economic Co-operation and Development (OECD), *Positive Adjustment Policies: Managing Structural Adjustment* (Paris, OECD, 1983).
3. See the Aluminum Association, *Aluminum Statistical Review for 1984* (Washington, D.C., Aluminum Association, 1985) p. 23.
4. The discussion in this paragraph and the next is developed from Merton J. Peck and John J. Beggs, "Energy Conservation in American Industry," in John C. Sawhill and Richard C. Cotton, eds. *Energy Conservation: Successes and Failures* (Washington, D.C., The Brookings Institution, 1986) pp. 86–90.
5. Maxine Savitz, "Energy Conservation Policies," in the Aluminum Association, *Aluminum Industry Energy Conservation Papers* (Washington, D.C., Aluminum Association, 1981) p. 19.
6. John A. Stuckey, *Vertical Integration and Joint Ventures in the Aluminum Industry* (Cambridge, Mass., and London, Harvard University Press, 1983). For a discussion of bauxite transactions, see p. 288; for alumina, see p. 292.
7. Organisation for Economic Co-operation and Development (OECD), *Aluminum Industry: Energy Aspects of Structural Change* (Paris, OECD, 1983) p. 99.
8. Ibid.
9. Ibid.
10. Calculated from *Metal Statistics 1984: The Purchasing Guide of the Metal Industries* (New York, Fairchild Publications, 1984) p. 24.
11. The terms *redfield, greenfield,* and *brownfield* come from the steel industry; they are easy to remember if the reader understands their origin. Redfield costs refer to existing plants, which in steel production light

up the horizon. Brownfield costs refer to the expansion of existing plants, presumably because adjoining land has been discolored by the steel furnace. The origin of the term greenfield for entirely new construction is obvious.

12. Based on table 7-3 in this volume.
13. Power costs are from Resource Strategies, Inc., "Electric Power Rates for the Aluminum Industry in the Pacific Northwest" (mimeographed report prepared for the Bonneville Power Administration, Portland, Ore., November 1984) p. 60.
14. Ibid., p. 50.
15. London Metals Exchange prices as reported in Bonneville Power Administration, *Selling Power: BPA's Direct Service Industries: Changing Conditions—Changing Needs?* (Portland, Ore., February 1985) p. 9.
16. For example, Sterling Brubaker finds that potlines "may be added at two-thirds to three-quarters the cost of new facilities and often more cells may be added to existing potlines at still less cost." (A potline is the unit of production consisting of a line of cells that reduce alumina to aluminum.) See Sterling Brubaker, *Trends in the World Aluminum Industry* (Baltimore, Md., Johns Hopkins Press for Resources for the Future, 1967) p. 95.
17. See Thomas Langton, *Investment Patterns in the United States Aluminum Industry, 1950–1977: A Historical and Empirical Analysis* (Doctoral dissertation, Pennsylvania State University, 1980) p. 92.
18. For a review of the studies on economics of scale in the primary aluminum industry, see Langton, *Investment Patterns*, pp. 42–44.
19. Calculated from data in the U.S. Bureau of Mines, "Primary Aluminum Plants World-wide" (mimeo, Washington, D.C., 1983).
20. Data from table 7-3 in this volume.
21. OECD, *Aluminum Industry*, p. 88.
22. Ibid.
23. For total capacity for market economies, see OECD, *Aluminum Industry*, p. 134; for estimated growth rates, see p. 56 of the same source. Production figures were taken from *Metal Statistics 1985* (New York, Fairchild Publications, 1985) p. 19; because the 1984 market economy production was equivalent to 1980 production, 1984 was used as the base year. The author then used these data to calculate which year production would equal capacity for the various growth rates.
24. See chapter 5 in this volume.
25. Ibid.

2

THE UNITED STATES: A TROUBLED INDUSTRY

MERTON J. PECK

The United States has always ranked first in both the consumption and the production of aluminum, accounting for approximately 25 percent of worldwide production and consumption. Yet 1985 production (3.5 million metric tons, or tonnes) was well below the industry's 1980 peak (4.7 million tonnes),[1] suggesting a troubled industry rather than a dominant giant. Initially, the industry's problems were blamed on the recession that began in 1979. Cyclical factors, however, may have only intensified the more deep-seated problem of the decline of the United States' international competitiveness as a location for aluminum smelting. This problem is the central theme of this chapter.

As noted in chapter 1, there are two kinds of international competitiveness or comparative advantage in the aluminum industry: (1) competitiveness for the location of new smelters and (2) competitiveness for the continued operation of existing smelters. There seems to be a popular view that the energy shocks of 1973 and 1979 eliminated U.S. competitiveness for the location of new smelters. But as this chapter will show, the international competitive position of the United States for new smelters began to decline as early as 1960, the year the last smelter was constructed in the United States by one of the "big three" U.S. producers (Alcoa, Reynolds, and Kaiser). Decline was gradual, however, rather than sudden because smaller producers continued to build smelters through the 1970s. The international search for low-cost power smelter sites that characterized the late 1970s was replicated within the United States almost from the beginning of the industry. Thus, the international stage is only the latest in a long-standing process of shifting locations to tap new sources of low-cost power.

The narrower and more complex question to be examined here is the international competitiveness of existing smelters, much of which depends on the cost of electricity. Electricity pricing in the United States is highly decentralized: power rates to smelters are individually set by the many suppliers of electricity, each of which may face a different cost-and-demand situation. As of 1985 some suppliers had power surpluses, in large part because they had built nuclear or steam generation capacity on the basis of forecasts of a high growth in demand and a low cost for nonhydro capacity, neither of which was borne out by subsequent developments. The whole issue of electricity rate setting for aluminum smelters is complex because of the uncertain duration of power surpluses; furthermore, the rates required to keep smelters operating competitively are difficult to determine and in constant flux because they must be tied to the state of the world market for aluminum.

Throughout this chapter the focus is on the big three U.S. producers—Alcoa, Reynolds, and Kaiser—which accounted for approximately 57 percent of U.S. primary aluminum capacity in 1984.[2] (By 1984 the big three had become the big four because Alumax's capacity exceeded that of Kaiser. We shall, however, continue to refer to the big three as including only the three older producers, the common use of the term until 1984.)

Like their international competitors, the big three have become increasingly transnational in their operations, although their production and sales are still predominantly in the United States. As of 1984, there were also eight other U.S. producers of primary aluminum, the outcome of successive entries into the market that began in 1951.[3] As noted in chapter 1, the shift from a highly concentrated to a less concentrated industry structure has characterized the aluminum industry worldwide, but only in the large U.S. market are there so many producers of primary aluminum. Since 1984 two of the smaller companies have dropped out of aluminum production, and the structure of the industry shows rising concentration. A return to the triopoly of 1951, however, is highly unlikely.

THE EXPANSION AND CONTRACTION OF THE U.S. ALUMINUM INDUSTRY

From 1950 to 1980 the U.S. aluminum industry was the leader in a worldwide expansion of production capacity to almost 4.5 million tonnes. The U.S. expansion started earlier than those of Europe and Japan and proceeded in two distinct waves—the first from 1950 to 1961 and the second from 1968 to 1973. All but three of the twenty-one smelters built after World War II were opened in one of these two periods (table 2-1).

The consumption of primary aluminum grew sharply at the beginning of each of the two expansion waves. The expansion of capacity that followed first lagged and then outpaced the growth in demand, a sequence that resulted in excess capacity by the end of each wave. Recessions, as in 1961 and even more markedly in 1976,

Table 2-1. Distribution of New Aluminum Smelters in the United States by Time Period, Region, and Power Source, 1950–1981

Region	Prior to 1950	1950–1955	1956–1961	1962–1967	1968–1973	1974–1981	Total plants in 1981
Southeast/Northeast							
Hydro	4	1	0	1	0	0	6
Coal	0	0	0	0	1	1	2
Natural gas	0	0	0	0	0	0	0
Total	4	1	0	1	1	1	8
Northwest							
Hydro	5	2	1	1	1	0	10
Coal	0	0	0	0	0	0	0
Natural gas	0	0	0	0	0	0	0
Total	5	2	1	1	1	0	10
Southwest							
Hydro	0	0	0	0	0	0	0
Coal	0	2	0	0	0	0	2
Natural gas	1	3	0	0	1	0	5
Total	1	5	0	0	1	0	7
Ohio Valley							
Hydro	0	0	0	0	1	0	1
Coal	0	0	3	0	2	0	5
Natural gas	0	0	0	0	0	0	0
Total	0	0	3	0	3	0	6
Overall total	10	8	4	2	6	1	31

Note: Some plants use more than one type of power. When a plant's power sources are mixed (some fraction of hydro and coal), the plant is credited to hydro because it is the low-cost hydro that dictated the location of the smelter. Similarly, when a plant uses both natural gas and coal, it is credited to natural gas because again it was the fuel source that set the location. One plant uses some oil, and four use some nuclear-generated electric power; but at none of these five plants is either oil or nuclear power the major source of electric power.

Source: Computed from data in U.S. Bureau of Mines, *Primary Aluminum Plants Worldwide 1982* (Washington, D.C., 1982).

not only stopped the growth in demand for aluminum but brought it down below its previous peaks, halting the expansion of capacity. Expansion occurred again only after demand resumed its growth.

Each expansion wave involved a shift in the geographic region in which smelters were built. The first wave made the Southwest a center of aluminum production, with natural gas the main source for generating electricity (see table 2-1). The last part of this wave brought the aluminum industry to the Ohio Valley to use coal to generate thermal power. The second wave, beginning in 1968, again relied largely on coal to generate electricity. Each shift, however, can be viewed as one segment of the ongoing search for low-cost power.

The First Wave: 1950–1961

The U.S. aluminum industry started the 1950s with a history quite different from that of most U.S. industries.[4] From its inception until 1940, there was only one producer of primary aluminum: Alcoa. Beginning in 1940, however, the U.S. government embarked on a program that expanded aluminum production capacity twofold over the next four years to meet the demands of the military. The government-owned Defense Plant Corporation built the new smelters, some of which were operated by Alcoa. Others were operated by newcomers. In bringing in new firms, the government was responding to antitrust action against Alcoa, which in 1944 resulted in a finding that Alcoa had monopolized the production of primary aluminum in the United States in violation of the Sherman Act. In 1946 the government-owned plants were first leased and then sold to Reynolds and Kaiser, creating the present-day big three producers. The sales, approved in 1950, brought to a conclusion the Alcoa antitrust case.[5]

Yet no sooner had the judge's ruling appeared to end government involvement in the aluminum industry than the Korean War led to further government-promoted expansion, again to meet national security needs. Capacity doubled once again, and the government program that produced the increase determined many of the characteristics of the industry's expansion in the 1950s.[6]

One of these characteristics was a reliance on natural gas-generated power. Prior to 1950 the conventional view was that only hydropower could be used to generate electricity for aluminum smelters; any other source would be too expensive to be competitive. In 1950, however, there was relatively little unused hydropower available in the United States. In addition, with the U.S. government insisting not only on a massive expansion but one that would bring smelters into early operation, building new dams would have been too lengthy a process. Thermal generating plants using natural gas as a fuel could be built quickly. Also, the cost of natural gas, although greater than hydropower, was still reasonable.[7] This was particularly true in the Southwest because there was limited demand for natural gas within the region, and the process of moving it by pipelines to the Midwest and Northeast was just beginning. (The price of natural gas in the early 1950s was between 5 and 10 cents as compared to approximately 90 cents in 1978.[8])

In response to the government's call for increased capacity, Alcoa expanded its one natural gas-supplied smelter and built another; Reynolds built two new smelters using electricity generated from natural gas. Three other smelters using hydropower were also built, one in the Tennessee Valley Authority area and two in the Bonneville area, which exhausted the hydro possibilities then available. (The Bonneville Power Administration [BPA] is a large federal agency supplying hydropower to the states of Washington, Oregon, Montana, Idaho, and California. The Tennessee Valley Authority [TVA] is a similar agency primarily serving Tennessee, Kentucky, Alabama, and, to a lesser extent, four other states.)

As the expansion continued, the opportunities to use cheap natural gas for aluminum smelting disappeared. The aluminum industry then turned to coal, building five smelters that relied on coal-fueled thermal generation of electricity. The five smelters were in the Ohio Valley and the Southwest. Again, coal generation was more expensive than hydropower, although the cost of coal-generated power was declining, reflecting technological advances in mining, coal transportation, and the efficiency of steam generating plants. Also, in the case of the Ohio Valley smelters, the savings in transportation costs that resulted from a location close to the East Coast fabricators offset in part the higher cost of coal-generated electricity.

During the 1950s expansion wave, three new companies entered the aluminum industry: Anaconda Copper Company; Harvey, an aluminum fabricator; and Ormet, a joint venture of the Revere Copper Company and the Olin Corporation. In themselves, these new entrants did not change the importance of the big three significantly, but they demonstrated that entry was feasible and established a pattern for even more entry in the next expansion wave.

The federal government promoted the expansion of the aluminum industry through a special set of incentives that reduced the risks of such expansion. For example, primary producers were awarded certificates of accelerated amortization, typically for 80 percent of the cost of building new plants or expanding existing ones.[9] Accelerated amortization was a commonly used device during World War II and the Korean War to encourage defense-related investment. It allowed producers to depreciate these costs in five years and resulted in depreciation rates that virtually eliminated any corporate tax liability for aluminum producers during the first five years of new smelter operation. Although the initial tax saving was offset by subsequently greater tax liability once the plants were fully depreciated, shifting the tax liability forward amounted to an interest-free loan from the government.

A more exceptional incentive was purchase contracts, which required the government to purchase at the published price all unsold aluminum from the new plants during their first five years of operation. Together with accelerated amortization, purchase contracts ensured that most of a producer's investment on a new plant would be recovered in five years, thus making the investment virtually risk free.

The government-sponsored program of capacity expansion ended in 1954, but its effects extended until 1961. The federal government's use of purchase contracts and accelerated amortization gave it control over the pace and direction of the expansion, which facilitated the industry's shift to higher cost power sources and the entry of new producers. The new plants built under the program were guaranteed a market by government purchase contracts; indeed, the government purchased substantial amounts for its national security stockpile in 1957.[10] Such purchases stabilized the overall market for aluminum and made the further expansion of capacity from 1954 to 1961 less risky.

The role of the federal government, however, was more to accelerate trends inherent in the development of the aluminum industry than to change the direction of the industry's growth. Even without the Korean War the demand for aluminum would have grown, and as the second expansion wave was soon to demonstrate, that demand would be accompanied by the entry of new producers and a shift toward thermal power. But it is accurate to assign to the government expansion program at least partial responsibility for the excess smelting capacity that marked the industry from 1959 to 1963 and brought the first expansion wave to an end.

The Second Wave: 1968–1973

The second wave of expansion began in 1968 and continued until 1973 in response to a growing demand for aluminum. Six new smelters were opened (see table 2-1), two using hydropower, three using coal, and one using natural gas. One of the hydropower-based smelters capitalized on additional BPA power, the other on the expansion of hydropower capacity in the Ohio Valley. The natural gas-based smelter was in Louisiana, a location decision that reflected the opening of new natural gas sources. These three smelters were the beneficiaries of the expansion of older sources of power; the construction of the three coal-based smelters reflected the lack of availability of hydropower.

Another characteristic of this second wave period was expansion through increases in the capacity of existing smelters, often a cheaper way of adding capacity than building new smelters, although it was limited by the availability of electric power (see chapter 1). During 1950–1961, 52 percent of the new capacity tonnage was in greenfield smelters, with twelve new smelters added. In contrast, in the 1968–1973 expansion, 80 percent was brownfield expansion—that is, adding capacity to existing smelters.[11]

The building of new smelters was limited to primary aluminum producers other than the big three. These smaller firms had fewer smelters from the previous expansion wave available for adding capacity; in addition, several were new entrants into primary production in the United States. By the mid-1970s there were twelve primary producers, an increase from three in 1950. Yet so much entry is a puzzle. Even though aluminum consumption was growing significantly, the rate of return on aluminum production was declining relative to that in other industries, a state of affairs that should have discouraged entry. Langton concludes that:

> The interest in aluminum production must be viewed from the standpoint of market growth. Aluminum consumption during the sixties expanded at approximately twice the rate of the economy. A considerable amount of enthusiasm for investment apparently was based upon potential long-run expectations due to future market expansion rather than profit rates which were based on historical trends.[12]

Yet even if one accepts an enthusiasm theory of investment, there remains the question of why these entrants did not build capacity elsewhere and ship primary aluminum into the United States. (This strategy was followed by the big three, for example, as discussed later.) A possible explanation is that companies believed they needed primary aluminum capacity within the United States to be effective participants in the U.S. market. In terms of delivery times and other customer services, the companies may have reasoned, a smelter abroad was not the complete equivalent of a smelter within the United States. The advantage of a U.S. location might have been necessary in their view to overcome the then-common belief that importers sold in the U.S. market only when demand was slack elsewhere and so were not reliable sources of primary aluminum. In contrast, the big three, with many smelters in the United States, could build smelters abroad without a loss in service and no risk of being seen as lacking a commitment to the U.S. market.

The primacy of low-cost power as a factor in locating aluminum smelters, then, appears to be qualified by a preference for market-oriented locations, particularly in siting the first few smelters of a firm. But it should also be noted that the second expansion wave took place when the United States—although not the lowest cost location—was still a country of moderately priced power.[13] Hence, the marketing advantages from a smelter located in the United States could be obtained without a great cost penalty. About this time there were also comparable waves of expansion in both Western Europe and Japan that similarly reflected the preference for market-oriented locations.

In this second wave of expansion the big three producers limited their U.S. expansion to adding capacity to existing smelters; their last greenfield smelter was opened in 1961. From 1961 to 1974, their percentage of capacity declined from 87 percent to 65 percent.[14] In the same period, the big three built twelve smelters in Europe, South America, and Australia. By 1982, U.S. firms had ownership interests in nineteen smelters outside the United States. Such interests were held predominantly by the big three; other firms were represented in only two or three such smelters.

The distribution of these nineteen smelters by time period, power source, and region is shown in table 2-2. The big three built two different types of smelters abroad: (1) those producing aluminum for local or regional markets to substitute for exports from the United States and (2) those producing aluminum for export to the United States or to other countries outside the region in which they were located. From 1962 to 1973, the smelters built in Western Europe served local markets—except in Norway. The two smelters located there were built with an export orientation, as were four

Table 2-2. Aluminum Smelter Capacity Outside the United States Constructed by U.S. Primary Aluminum Producers

Period	Capacity (thousands of tons)	Distribution by power source (percent of capacity)			Distribution by country (percent of capacity)			
		Hydro	Coal	Other	Australia	Brazil	Canada	Other
1955–1961	199	100	0	0	0	0	80	20
1962–1967	526	76	24	0	24	0	0	76
1968–1973	351	37	48	15	0	0	13	87
1974–1982	296	32	68	0	69	21	0	10

Note: Capacity reflects both initial construction and subsequent expansion of the plant. It is assigned to time periods by the date of initial operation of the plant.

Source: Compiled from data in U.S. Bureau of Mines, "Aluminum Plants Worldwide" (mimeo, Washington, D.C., 1982).

others—one each in Canada, Surinam, Venezuela, and Ghana. All six of the export-oriented smelters used hydropower; indeed, the availability of cheap hydropower explains their location. In contrast, locally oriented smelters had a variety of power sources, with coal and nuclear power of significant importance.

U.S. firms built smelters abroad mostly through joint ventures, with only four of the nineteen smelters owned entirely by a single U.S. producer. The participants in these joint ventures varied from governments and local companies to private investors to other U.S. primary producers. Sharing the ownership of a smelter allowed an aluminum producer to diversify the risk associated with one smelter by distributing the investment over several smelters. Such risk sharing also occurred in the United States but only among firms outside the big three. The difference in the behavior of the big three at home and abroad may reflect in part the greater risk of investment abroad, the absence of antitrust constraints, and the preference of governments for equity participation either directly or by local companies and investors.[15]

The investment behavior of the big three during the second wave of expansion of aluminum production capacity thus demonstrated the decline of the United States as a competitive location for new smelters even before the first oil shock in 1973. These companies chose to build smelters abroad; they had the ability to search out locations worldwide, build the necessary infrastructure, and negotiate with local governments, all of which made their investment behavior a good indicator of international competitiveness.

The U.S. Aluminum Industry After 1975: Recession and Contraction

The chronological story of the industry resumes in 1975 with the sharp downturn in the consumption of aluminum that occurred that year, reflecting the general recession in the U.S. economy. Although consumption recovered through 1979, the increase was insufficient to reach the 1974 level or to match the capacity levels created in the second wave. In 1981 consumption again declined and remained at low levels during 1982 and 1983. It recovered somewhat in 1984 but declined again in 1985. Not only is it apparent that the second wave of expansion has ended, it also seems unlikely that there will be a third wave.

Table 2-3 shows U.S. aluminum production, imports, and inventory for three years: 1979, the best post-1975 year; 1982, a very bad year; and 1984, a year of modest recovery. Secondary aluminum production increased during this period, reflecting the factors discussed in chapter 1 and in particular the greater use of aluminum in beverage cans, which have a short life and a high recovery rate. Net aluminum imports increased sharply, with most of the increase coming from Canada. By exporting extensively, the Canadian producer Alcan was able to maintain a 90 percent capacity utilization rate throughout the recession. In contrast, in the United States, output fell by 28 percent, and capacity utilization was 60 percent by the end of 1982.[16] To maintain its levels of output, Alcan shaded prices, that is, it sold at prices below its published prices. It was possible to do this because Alcan's reliance on hydropower meant that its variable costs were exceptionally low. Thus, as discussed in chapter 1, sales at low prices would still have covered variable costs and hence would have been profitable.

Table 2-3. Changes in U.S. Aluminum Production, Imports, and Inventory for Selected Years, 1979–1984 (thousands of short tons)

Item	1979	1982	1984
Production			
Primary	5,023	3,603	4,518
Secondary	1,777	1,836	1,940
Net imports	+13	+278	+686
Inventory change	−184	−203	+427

Source: Compiled from data in the Aluminum Association, *Aluminum Statistical Review, 1984* (Washington, D.C., 1985) p. 61. Inventories refer to producer stocks.

Alcan is not the only non-U.S. producer that has been able to preserve a high capacity utilization rate. A pattern of aluminum industries outside the United States maintaining production has led to the concern that U.S. smelters have become the worldwide swing capacity—that is, the capacity that goes in and out of production as worldwide demand for aluminum fluctuates. That role would mean employment instability for the U.S. smelting industry, as well as reduced incentive to modernize smelters that are used only periodically. These concerns are a central element in U.S. policy toward the industry, which is discussed in the next section.

The problems of the U.S. aluminum industry, however, are not merely cyclical. Of the thirty-one smelters operating in 1981, five were not operating in January 1985.[17] These five appear to be the high-cost smelters, particularly the three that rely on natural gas-generated electricity, a fuel whose price increased more than five-fold from 1970 to 1978 and whose supply to smelters was subject to regulatory curtailments.[18] Of the remaining smelters that closed, one was served by TVA and the other by BPA. (Smelter closings are not announced by companies as such; rather, in keeping with common practice, the discussion here treats the cessation of production as a smelter closing.)

ELECTRICITY COSTS AND INTERNATIONAL COMPETITIVENESS

Competitiveness for New Smelters

The big three, as discussed in the previous section, have not built a single new smelter in the United States since 1961, an indication that the United States began losing its international competitiveness some time ago. The extent of the loss was modest in the 1960s and became marked only after the first oil shock. Notably, only one smelter, at Mount Holly, South Carolina, came into operation after 1974. This smelter used a large block of electric power that became available through the construction of large-scale fossil fuel and nuclear power facilities.

The revealed pattern of actual investment in new plants is consistent with the results of a World Bank study (table 2-4). Decisions about where to locate new smelters predominantly depend on electricity costs, and more on future than current electricity costs because the life of a smelter is thirty or more years. As the table shows,

Table 2-4. Aluminum Production Costs at New Plants at Selected Sites in the Year 2000
(1980 U.S. dollars)

Country	Total costs (per tonne)	Power costs (per tonne)	Total costs (cents per pound)
United States	2,260	630	102
Japan	2,270	630	103
Western Europe	2,310	630	105
Australia	1,870	270	85
Brazil	1,860	270	84
Canada	2,030	400	92

Source: From Martin Brown and coauthors, *Worldwide Investment Analysis: The Case of Aluminum* (World Bank Staff Working Paper no. 603) (Washington, D.C., 1983) reprinted by permission of the World Bank. The costs are in 1980 dollars for plants built around the year 2000, but they differ from estimates of current costs only in assuming a rise in real thermal power costs at an annual rate of 3.2 percent, mitigated by an annual improvement in energy efficiency of 0.5 percent.

the United States, together with Japan and Western Europe, are projected to be high-cost locations relative to the ABC countries (Australia, Brazil, and Canada). Of course, the major factor in the cost difference is the variation in the costs of electric power in the six countries. (The electricity costs shown in the table reflect the fact that additional power in the United States, Japan, and Western Europe must come from high-cost thermal plants.) The World Bank study calculates, for the most likely increases in demand, the pattern of new smelter location that would minimize the worldwide cost of producing aluminum. The study projects that only three new smelters—3 percent of the worldwide increase in capacity—will be built in the United States between 1980–2000. Yet even this modest increment of new capacity now seems unlikely.

International Competitiveness for the Operation of Existing Plants

Existing smelters represent capital costs that have already been incurred; the decision to operate an existing smelter is based largely on whether the price of aluminum equals or exceeds the unit operating or variable costs. The closing of some U.S. smelters and the partial utilization of others, however, raises the question of whether some significant fraction of existing U.S. smelters is no longer internationally competitive.

In analyzing this question, the smelters already closed by 1984 can be set aside; presumably, these were high-cost operations. What is of interest are the twenty-six U.S. smelters operating in 1986.

As might be expected from the historical development of smelters located in different regions and with different sources of power, the twenty-six smelters pay quite different prices for electricity. Table 2-5 arranges the U.S. smelters by the estimated 1983 electricity cost per pound of aluminum—that is, the kilowatt-hour (kWh) consumption per pound times the price per kWh. Thirty-seven percent of the U.S. smelter capacity is in the high-cost category—those smelters with power costs of over 20 cents per pound of aluminum—and it is this fraction of U.S. capacity that is at risk in two respects.[19]

The first is the risk of becoming permanent swing capacity. The fluctuations that have occurred in the demand for aluminum have resulted in variations in annual aluminum production of approximately 30 percent since 1975.[20] High-cost plants are the obvious candidates for output cutbacks during periods of low demand, and, indeed, that has been their recent history. For example, as of January 1, 1985, nineteen of the twenty-six smelters were operating at 100 percent of their rated capacity. The remaining seven were substantially under-utilized, with the percentage of capacity utilization as low as 13 percent in some cases. Given the large variable cost differences among smelters, the industry practice is to close high-cost smelters or cut back their production sharply during demand downturns while continuing to operate low-cost smelters at full capacity.

Table 2-5. U.S. Smelting Capacity by Estimated Electricity Costs per Pound of Aluminum, 1983

Cost	Annual capacity (in short tons)	Percentage
Less than 15 cents	1,204,000	26
15 Cents to 20 cents	1,707,000	37
Over 20 cents	1,708,400	37
Total	4,619,400	100

Note: Data were unavailable for six smelters listed as in existence as of January 1, 1984. Of these six, three were not operating for most of 1983, and data were unavailable for the three others. The smelters included in the above computation account for 92 percent of the rated U.S. capacity operating in 1983. The calculation reflects both the price of electricity and variations in smelter efficiency in using electricity.

Source: Industry data supplied to the author.

The second risk for these high-cost smelters is permanent closing, an outcome that is increasingly appealing to producers as low-cost capacity comes into operation in the ABC countries. Given electricity prices of as low as 15 mills, new smelters in these countries are likely to have total costs (including capital costs) that are well below the average variable costs—70 cents—of U.S. high-cost smelters. Aluminum companies thus would benefit from increasing their capacity abroad and closing their high-cost smelters in the United States. That process, however, will be related to demand cycles in aluminum production. On the upswing, new capacity will come into operation; on the downswing, high-cost smelters will close, at first temporarily and then permanently.

All but one of the high-cost smelters are in the TVA and Bonneville areas. These two regions initially were low-cost power areas when the smelters were built, and they are still substantially dependent on hydropower, the traditional low-cost power source. But as table 2-6 indicates, they are now areas of high-priced electricity. In both regions, economic growth led to growth in the demand for electricity; that demand exceeded the area's low-cost hydro capacity and led to the construction of high-cost nuclear or coal generating capacity. The change to high-cost power sources sharply increased the electricity rates charged to aluminum smelters in these two regions in the 1970s and early 1980s. As a result, smelters that had originally been low-cost operations became high-cost facilities.

Table 2-6. Estimated 1983 Power Costs for Aluminum Production

Region	Mills/kWh
Northwest (BPA)	25.9
Southeast (TVA)	30.2
Ohio Valley	24.6
Gulf Coast	19.6
Northeast	10.3

Source: "The Direct Service Industries and Their Role in the Regional Economy, A Report by Public Utility District No. 1 of Snohomish County" (September 25, 1985) p. 14. There have been some reductions in power since then, most notably in the Northwest. The BPA rate was changed to 22.8 in July 1985; in addition, there was a 5-mill discount available beginning in September 1986 for limited blocks of power. The low rate for the Ohio Valley reflects long-term power contracts which will expire in a few years.

The high-cost U.S. smelters also tend to be old—all but one were built more than twenty-five years ago.[21] Old smelters have a high consumption of kilowatts per pound of electricity. It has been estimated that the Alcoa Vancouver, Washington, plant built in 1940 uses 8.85 kWh per pound of aluminum produced, whereas the Conalco plant in Goldendale, Washington, built in 1970, uses 6.35.[22]

The high-cost plants, then, combine a location in a high-priced power area with older technology—and, in the case of the plants in the Northwest, high transportation costs to East Coast and Midwestern markets. Unlike people, aluminum plants can be made young again but only with substantial expenditures on modernization. Primary producers have been reluctant to modernize their high-cost plants unless they obtain favorable rates for electric power. Such rates, however, are not totally a fact of nature; they are largely set by public policy, as discussed in the next section.

PUBLIC POLICY

It is often said that the United States is a nation without an industrial policy—that is, an explicit policy for the problems of individual industries. But if the definition is broadened to include ad hoc measures, then the United States does in fact have an industrial policy. The steel and automobile industries are the most visible examples of cases in which ad hoc government actions—import quotas, loan guarantees, and regulatory changes[23]—have been devised to help industries in trouble.

In the last decade the U.S. aluminum industry has encountered problems comparable to those of the steel and automobile industries. There has been import competition, company losses, layoffs, low capacity utilization, and little new investment. The aluminum industry also faces the slower growth in demand typical of a mature industry and the loss of international competitiveness. Federal policy measures to provide some relief for the aluminum industry can be summed up in one word: inaction.

The policies that have, in fact, mattered for the aluminum industry are those adopted by the suppliers of power. Inaction is not an option for them; they must set electricity rates for aluminum smelters, and those rates determine the international competitiveness of the U.S. industry. Indeed, as chapter 1 indicates, electricity rates are perhaps the most important policy instrument that can be used to promote or maintain the aluminum industry.

In the United States the supplying of power is highly decentralized and diverse. Suppliers of power to the aluminum industry include regulated investor-owned utilities, regional cooperatives, and state power organizations; the two big federal power entities, TVA and BPA; and generating facilities owned by the aluminum companies. Apart from the last, every supplier is either regulated or owned by the government. The regulation, however, is by state commissions, and each government-owned power supplier pursues a largely independent policy. In this respect the United States is like Canada, Australia, and the Federal Republic of Germany, nations in which state or provincial governments are as significant as the federal government in determining the fate of the national aluminum industry.

Traditionally, the various suppliers have sold electricity to aluminum producers under long-term contracts that guaranteed lower rates than those offered to other consumers. These contracts expired in the late 1970s; since then, the rates to smelters have increased sharply. The extent and character of the increases have varied among regions, of which the most important have been the Northwest, served by BPA, and the Southern states, served by TVA.

The Bonneville Power Authority

Bonneville provides power for a third of the smelting capacity in the United States;[24] the smelters in their turn accounted for about 32 percent of BPA's 1984 electric power revenues.[25] No other power supplier is as important to the aluminum industry, and few others have the same dependence on smelters as customers. Even though BPA is a federal agency, its policies largely are determined independently of the Department of Energy, of which it is a part. Electricity rates, for example, are set by the BPA administrator, although that authority is limited by legislation (in particular, the Pacific Northwest Electric Power and Conservation Act of 1980).

Bonneville's current situation (as of 1985) had its origins in the hydro capacity created by the New Deal policy of building large-scale dams as part of a strategy of regional development. By 1940, BPA had a large surplus of power, which it sold at only 3 mills/kWh. In response to such low rates, aluminum smelters came to the Northwest beginning in 1940.[26] The revenues from power sales to smelters enabled BPA to build its transmission grid, round out its dam system, and make payments to the federal treasury for interest on the government's investment. The availability of low-cost power continued to attract smelters until as late as 1970—although after 1950 there was never enough cheap BPA power to serve all the smelters that sought such power. By 1970, BPA was serving ten of the thirty U.S. smelters then operating or under construction.[27]

In June 1976, BPA notified its aluminum customers that current contracts providing 3-mill power would not be renewed when they expired between 1984 and 1987.[28] This action reflected a forecast that the demand for electricity from BPA's other customers would increase sizably—so much so that it would require all of the existing capacity to meet that demand. With no plans for further expansion of hydro capacity, BPA's major source of power at that time, BPA could not continue to supply cheap power to the smelters. Indeed, the forecasts of demand growth for the Northwest were so sizable that BPA acquired much of the projected output of the planned nuclear plants of the Washington Public Power Supply System (WPPSS).[29] Although cost estimates for the nuclear-generated power turned out to be higher than anticipated, it was clear from the outset that nuclear-generated power would be higher cost than BPA hydropower.

BPA's actions of June 1976, combined with earlier notification that BPA would be unable to sell power to investor-owned utilities after 1973, created what has been called the Northwest power crisis. The disappearance of unused hydropower raised BPA rates, bringing particularly large increases to the smelters (rates increased by approximately ninefold from 1978 to 1983). The Northwest became second only to the TVA area as a region of high-priced power for aluminum smelters.

These rate increases were premised on forecasts that the demand for power in the Northwest would exceed BPA capacity at an early date. Like many forecasts made at that time, these turned out to be wrong for two reasons: (1) higher prices for power cut back sharply the growth in electricity consumption, and (2) the growth rate of the U.S. economy fell markedly in the 1970s, which in turn reduced the growth of power consumption. By the summer of 1984, BPA found itself with a power surplus and offered aluminum smelters a rate of 22.7 mills for six months; in March, rates were reduced again to yield an average rate of 19.8 mills.[30] Yet, even with these reductions, Bonneville had a critical problem over the long run in competing with 15-mill power in Canada and Brazil.

With a power surplus, the Northwest power situation in 1985 assumed a quite different character than had been anticipated. Instead of allocating electricity among eager customers, BPA found itself working to maintain its power sales. In hindsight, the construction of the nuclear power plants, which added to the surplus, was a clear mistake. Not only were they unneeded but their construction costs, which were estimated in 1977 at $6.6 billion, had risen and were estimated in 1983 to be $24 billion to complete the plants.[31] (Errors of this magnitude in the cost estimates for nuclear plants were common throughout the world; see, for example, chapter 3.) Although two of the WPPSS nuclear plants were cancelled, one is operating and two others are 60 to 75 percent complete with construction temporarily suspended.[32] But even if the two plants were to be cancelled, BPA 1985 forecasts would still show a power surplus until 1991.[33] If the high-cost aluminum smelters using BPA power were to close, the surplus is forecast to continue until about 1995.[34] Under alternative assumptions, the power surplus has been forecast to end as early as 1990 or to persist beyond 2005.[35] It is hard to view such predictions without a good deal of skepticism, however; as this account suggests, forecasts of the demand for power have been notoriously inaccurate.

In June 1985, in response to the power surplus, BPA announced that it was considering three options:[36]

1. A long-term variable electricity rate that would fluctuate with the price of aluminum. When the price of aluminum was low, the smelters would receive rate discounts; when the price was high, the smelters would pay rate premiums. The plan's objective was to keep the BPA smelters operating throughout the fluctuating cycles in aluminum demand by varying smelter operating costs (within limits) to keep such costs below the market price of aluminum.
2. A conservation/modernization program for the Northwest smelters. The objective of this option was to reduce the costs of the smelters—by reducing electricity consumption through efficiency gains—so that they could be more competitive internationally. BPA was to finance less than 20 percent of this program; the aluminum industry would finance the rest, with possible concessions from state and local governments. An extensive modernization program for all ten smelters was estimated to require an investment of $320 million.
3. The establishment of a long-term linkage between industrial and utility rates. The goal here was to provide a more predictable industrial rate to facilitate long-term planning by the aluminum companies.

These proposals were subject to lively debate, a reflection of the great uncertainties involved and the major consequences of each option for different BPA consumers. One set of uncertainties involved the duration of the power surplus. If the power surplus disappeared before the rate concessions to the aluminum smelters expired, rates to other power customers would rise to offset the concessions to smelters; also, the availability of power to these customers would be reduced. If the power surplus continued and the rates were raised in anticipation of a shortage, losing the demand represented by the smelters would raise rates to other consumers. Another set of uncertainties related to the size of the rate reductions—how large they would have to be to keep the BPA smelters operating—and their most effective form. Still another group of uncertainties involved the price at which BPA would have been able to sell its surplus power to California, a strategy that could provide an alternative source of demand to the smelters. A final set of uncertainties involved the value of the smelters to the economy of the region.

Choosing the final package was a highly politicized process involving a number of well-organized groups: the aluminum companies, the unions representing aluminum workers, local governments, the public utility districts, and investor-owned utilities. One of the most striking features of the debate surrounding the decision making was that it was conducted strictly in terms of regional interests. One objective was to stabilize and increase BPA revenues so BPA could make the required U.S. Treasury payments on past federal investments;[37] another was to keep rates low to other BPA customers. A final objective was to maintain the smelters' contributions to the regional economy and employment. The national interest in the aluminum industry, on the other hand, received only an occasional sentence.

An interim step was taken in July 1985 when BPA announced an optional industrial incentive rate. This proposed rate was a 5-mill discount from the current industrial rate of 22 mills, creating a low 17-mill rate. The incentive rate was available for nine months beginning in September 1985 when the demand from other customers was expected to be low. Industrial customers were required to make definite quantity commitments to receive the incentive rates under a take-or-pay provision.[38] Yet the incentive rates were attractive only to some of the smelters, for as of January 1985, six BPA smelters were operating at full capacity; one was operating at 81 percent capacity; two were operating, respectively, at 50 percent and 47 percent capacity; and one was closed.[39]

In early 1986, BPA announced a longer run plan, which tied the price of electricity offered to smelters to the world price of primary aluminum. With a world aluminum price of 58 cents or less per pound, the electricity rate would be 14 mills; with a price of 59 to 72 cents, the rate would be 22.8 mills; over 72 cents, the rate would be 29.6 mills. When aluminum prices were high, the producers would pay about 7 mills more than their current rates; when prices were low, they would pay 7 mills less. For their part the aluminum producers had to commit themselves to the plan for five years with an option to renew for another five. The plan also provided an incentive for modernization and energy conservation by offering a 5-mill discount for every kilowatt-hour improvement in the efficiency of electricity use.[40] By July 1986, all the primary producers in the Northwest had agreed to the new BPA rates.[41]

The Tennessee Valley Authority

TVA's position—serving four smelters—was quite different from that of BPA in the early 1970s. TVA generated less than 20 percent of its power from hydro sources and relied mainly on coal-generated power; BPA, on the other hand, was totally dependent on hydropower.[42] In addition, TVA was more affected by the first oil shock because coal prices also increased sharply. Thus, the authority's rates began their sharp increase in 1973, whereas BPA rates increased some six years later. Finally, TVA was and is less dependent on aluminum smelters for sales of electric power. As a result, it seems more reconciled to the possibility of the loss of its aluminum customers.

In the 1980s, TVA, like BPA, developed a power surplus. In November 1983, the authority offered "interruptible" power (power that can be withdrawn if it is needed for other customers) on a one-time basis to industrial customers at a 25 percent discount from its 36-mill rate and with a ten-year contract. Of the four producers served by TVA, only Alcoa and Reynolds took such contracts; Revere and Consolidated did not.[43] As of January 1985 the Alcoa and Reynolds smelters were operating at full capacity, the Revere smelter was closed, and the Consolidated smelter was operating at 15 percent of capacity.[44]

Other Areas

Approximately half of the U.S. primary aluminum capacity is served by about twenty different power suppliers.[45] The focus here is on two of the largest: the New York State Power Authority and the Big Rivers Electric Corporation.

New York State provides hydropower from Niagara Falls and the St. Lawrence River to an Alcoa and a Reynolds smelter. Under a 1981 long-term contract the rate to these smelters was 13 mills, the lowest rate for purchased power paid by any U.S. aluminum smelter.[46] That rate was challenged, and the Federal Energy Regulatory Commission found that the allocation of hydropower by the authority discriminated against municipal consumers in favor of industrial consumers.[47] In 1983 the governor established the Temporary Commission on the Allocation of Power Authority Hydroelectric Power, which concluded the following: "Recently renewed contracts to supply low cost hydropower to Alcoa may be subject to claims of illegality in light of existing State preference requirements (for municipal power);" but "the Commission believes these contracts should not be disturbed to preserve the economy of the State as a whole." It was also the commission's view that the "potential economic benefits to the people of the State resulting from creation (and retention) of jobs outweigh the benefits resulting from moderate reduction in residential electric bills."[48]

The pursuit of this policy, along with the assurance of low-cost hydropower, will probably make these two smelters internationally competitive for some time to come. What is more uncertain is whether the policy is socially optimal or economically efficient; data are lacking for these judgments.

The Big Rivers Electric Corporation is a Kentucky utility owned by 931 rural electric cooperatives. In 1983, it supplied power to two Kentucky smelters—one owned by National-Southwire and one owned by Arco—which accounted for 70 percent of the utility's load.[49] Given that dependence, the closing of either smelter would threaten the financial viability of the utility.

In the late 1970s, Big Rivers, along with most utilities, forecast a big increase in consumption and in 1980 began construction of a large, $750 million coal-fired generating plant with financing from federal loans to the cooperatives. The capacity of the plant, which is now partially completed, turned out to be unnecessary for the actual power demands of the early 1980s and for currently projected demand. In the spring of 1984, Big Rivers applied to the state utility commission to include the plant in its rate base, an application that was subsequently withdrawn when National-Southwire objected.[50] (In the summer of 1984, Arco sold the other smelter served by Big Rivers to Alcan; Alcan's plans for the smelter are uncertain.) As of 1985, the Big Rivers dispute remained unresolved. In February 1985 the U.S. Department of Justice brought a suit to foreclose on the $1.1 billion loan that was used to build the steam plant.[51]

Welfare Economics and the Criteria for Electric Rates

As discussed in chapter 1, the economic criterion for pricing is that prices should equal marginal costs. To justify the production of the last unit of output, consumers should value that unit sufficiently to pay the costs to society of producing it.

Applying this apparently simple principle to electric power is difficult, however. If there is a power surplus, then the marginal cost is low because the marginal cost is only the variable costs of using the capacity to generate electricity rather than letting it stand idle. Variable costs are close to zero for hydro capacity, but even for high-cost nuclear plants, variable costs are still approximately 9 mills/kWh. If, on the other hand, there is a power shortage such that incremental demand must be served by building nuclear power capacity, the marginal cost can be 60 or 80 mills or even higher.

One seemingly easy way to resolve the uncertainty as to marginal cost is to set a rate that reflects the existing power situation: a low rate with a power surplus and a high rate when there is a shortage. Yet if the power surplus is of short duration, a low rate can provide misleading signals if consumers (and particularly smelters) make costly investment and consumption decisions based on the assumption of a continued low rate. Power customers can be adequately forewarned of the temporary character of a low rate, but even so there may be political pressures to continue the low rate after the power shortage ceases.

All of this, of course, assumes that the duration of a power surplus can be accurately forecast; recent U.S. experience, however, shows it cannot—at least, not with present forecasting tools. As indicated in the discussion of BPA policy, uncertainty makes the choice of power rates for aluminum smelters a very difficult decision.

The marginal cost criterion is only part of the overall welfare standard. The other part is to set some or all of the rates sufficiently above variable costs to collectively cover the capital costs of generating power. Profit maximization dictates price discrimination—high rates for users with inelastic demand and low rates for users with elastic demand. Profit maximization is not the only consideration, however; there are also issues of equity among consumers. Given the monopoly position of most electricity suppliers, public policy has not allowed them to set unregulated monopoly prices. Equity issues also exist between investors and consumers, in particular because regulation has traditionally limited the returns of monopolistic suppliers.

The capital cost issue recently has taken the form of who should pay for the excess utility capacity that has resulted from the optimistic forecasts of electricity demand—the consumers or the investors—and if it is to be the consumers, which group of them. The issue is still unresolved for many utilities but generally tends to be settled by compromises that divide the losses among all involved groups.

As long as power surpluses exist—which for BPA could be a considerable time—aluminum producers can expect low rates; but most U.S. power suppliers are unlikely to offer the long-term commitments to low rates that are necessary to prevent more smelter closings or keep many smelters from remaining swing capacity. In addition, the economic claims and political power of other power customers are sufficiently strong to preclude low rates in periods of power shortages. Consequently, as power demand grows in the United States, the nation will gradually lose its competitiveness even for the operation of existing smelters.

Policies of the Federal Government

The federal government has sole responsibility for trade and tax policies and has used these measures, as well as occasional loan guarantees, to help industries in trouble—for example, the automobile and steel industries. The federal government has not instituted such measures for the aluminum industry despite its recent troubles, and that is because the magnitude of the aluminum industry's problems is significantly less than for automobiles and steel.

Major losses in employment and financial crises in major companies are the events that have produced federal policy responses. Because aluminum smelting is not labor intensive, the recent decline in aluminum production has contributed relatively little to overall unemployment. For example, from 1979 to 1983, employment in primary aluminum smelting declined from 36,800 to 26,100—not a very impressive loss of jobs compared to the loss from one major steel or automobile plant closing. In contrast, automobile industry employment declined from 990,400 to 767,800 from 1979 to 1983, and steel industry employment went from 570,500 to 343,100. The decline in aluminum employment is not trivial to those who lost their jobs, but its magnitude is unlikely to trigger the type of policy response that would result from more massive unemployment.[52]

Financially, the individual aluminum companies in the late 1970s and early 1980s have been much better off than steel and automobile producers. None of the aluminum producers has experienced the financial crisis of a Chrysler even though profits in aluminum fell sharply with the decline in aluminum consumption. In 1979, Alcoa reported a net income of $504.6 million, which was the highest in the company's history. In the 1980s, Alcoa's net income was $174 million in 1983 and $256 million in 1984; the company suffered a loss of $17 million in 1985.[53]

Alcoa derives 97 percent of its revenue from aluminum operations, but most of that revenue comes from the sale of semifabricated products: sheet, extrusions, coils, and foil. Only 11 percent of Alcoa's revenue comes from the sale of primary aluminum. Semifabricated products are differentiated products in which service, research and development, advertising, and marketing are factors in market success. Fabricated products have been less subject to price cutting than primary aluminum, and the greater price stability of fabrications has moderated the fall in Alcoa's profits. Nonetheless, the profit decline has been a shock to a company that has been almost consistently profitable.

Reynolds and Kaiser have suffered more than Alcoa. In 1979, Reynolds's net income was $177.1 million. In the 1980s, Reynolds's net income was a loss of $99 million in 1983, a profit of $137 million in 1984, and a loss of $292 million in 1985. Kaiser followed approximately the same pattern, although it should be noted that at Kaiser a chemical division accounts for 30 percent of the company's revenues and the financial results reflect both the chemical and aluminum division operations. In 1979, Kaiser's profits were $232.2 million. In 1983, however, the company suffered a loss of $50 million; in 1984, a loss of $97 million; and in 1985, a loss of $187 million. Although these developments reflect an industry in trouble, the losses are still modest compared to either the automobile or steel industries during their most extreme difficulties. For example, all of the big three automobile companies (General Motors, Ford, and Chrysler) reported losses that were fourfold those of the aluminum industry for at least one year between 1979 and 1983; Chrysler alone had losses in three of the five years ranging from $476 million to $1.7 billion. Three of the bigger steel companies (U.S. Steel, Bethlehem Steel, and National Steel) recorded at least one year of losses in that period that were in excess of the highest loss of the big three aluminum companies.[54] By comparison with these two industries, the financial difficulties of the big three aluminum producers are more modest, as are the job losses.

The fortunes of the smaller producers during this period varied, and given their multiproduct character the financial viability of aluminum production is better reflected in their decisions to leave the aluminum business than in their company profits. Martin Marietta has sold one smelter to Comalco, Ltd., of Australia, and its second smelter has been closed. Revere Copper and Brass has closed its one wholly owned smelter although it still has an interest in the one smelter owned by the Ormet Corporation, a joint venture of Revere and Consolidated. As noted earlier, Arco has sold its one smelter.[55]

If the policy process is viewed as one in which only large and acute problems can attract attention, then the relatively modest problems of the aluminum industry would explain the lack of a policy response. But there may be still another factor: the big three have protected themselves against the loss of international competitiveness by constructing smelters abroad. This strategy is in contrast to that of the U.S. steel industry, which has invested little in the steel industry abroad. And although the U.S. automobile industry has had large operations abroad, until recently these have been quite separate from domestic operations, producing different models that were not exported to the United States.

Many of the initial investments of the big three producers in smelters abroad were, like the automobile industry, to serve their foreign markets. But foreign investment is now dominated by the search for low-cost power, and smelters built in the ABC countries will be serving the U.S. market. Most of those smelters will be partially owned and operated by U.S. primary producers; their costs will be lower than those of U.S. smelters, and U.S. producers will increasingly rely on such offshore operations. This gives U.S. aluminum producers an interest in free trade, an interest that is perhaps unique in U.S. industry. Free trade also serves the interest of the aluminum fabrication industry, which has twice the number of jobs as smelting.

THE U.S. ALUMINUM INDUSTRY OVER THE LONG TERM

The high cost of electric power in the United States has meant that this country has lost its international competitiveness for the construction of new smelters. In addition, the pattern of actual investment has conformed to the economic realities of relative costs.

There is no evidence that the recent pattern will change, although an occasional smelter may be built to take advantage of a large block of power that becomes available—for example, if and when a large nuclear or fossil fuel plant is constructed.

This conclusion holds despite the pockets of power surpluses now found in such suppliers as BPA or the Big Rivers Corporation. Temporarily, the surpluses have produced low rates, but these surpluses result from the overexpansion of high-cost nuclear and fossil fuel plants. The low rates reflect the existence of excess capacity and not power that is cheap to generate. Such rates are not a solid basis for constructing new smelters.

Nor in the long run are they the basis for the continued operation of existing high-cost smelters because economic growth, even at a slow rate, will continue to provide alternative markets for electric power with consumers that can outbid the smelters for power. And, as noted earlier, there is little indication that the smelters will be able to obtain low-cost power once the current power surpluses disappear. Like Japan, the United States seems to be becoming a major aluminum importer, although the process may take two or three decades rather than one.

An increasing reliance on aluminum imports might have raised concerns about national security, but it has not. U.S. aluminum production has for most of its history depended on imported bauxite and alumina; greater reliance on imported aluminum adds little to the already great dependence of the U.S. economy on the rest of the world. Aluminum is still one of the metals for which there is a stockpile objective, but national security concerns of this type seem dated in a world of ballistic missiles and have played no role in recent national security debates.[56] This attitude is in sharp contrast to the priority attached to an adequate supply of aluminum for defense in the 1940s and 1950s.

There are then no major interests that seem likely to promote a policy that would restrict the shift of primary aluminum smelting abroad. The speed of that shift will be determined largely by relative cost-and-demand conditions for primary aluminum and by public policy in the countries that are now the low-cost locations for new aluminum smelters.

NOTES

1. Data from the Aluminum Association, *Aluminum Statistical Review, 1985* (Washington, D.C., 1986) p. 45.

2. Ibid., p. 26.
3. In order of capacity the firms were Alumax, Arco, Consolidated, Martin Marietta Aluminum, Noranda, Revere, National Aluminum, and Southwire (ibid., p. 26).
4. The history of the aluminum industry, including the expansion wave of the 1950s, is discussed in detail in Merton J. Peck, *Competition in the Aluminum Industry, 1945–1958* (Cambridge, Mass., Harvard University Press, 1958); Thomas Langton, *Investment Patterns in the United States Aluminum Industry, 1950–1977: A Historical and Empirical Analysis* (Doctoral dissertation, Pennsylvania State University, 1980); and U.S. Department of Commerce, *Materials Survey on Aluminum: Facilities Expansion in the United States, 1940–1954* (Washington, D.C., 1956).
5. Alcoa was also allowed to acquire some facilities that were located at its existing plants.
6. Peck, *Competition in the Aluminum Industry,* pp. 150–152.
7. U.S. Department of Commerce, *Materials Survey,* p. VII-13.
8. Crawford D. Goodwin, ed., *Energy Policy in Perspective: Today's Problems, Yesterday's Solutions* (Washington, D.C., The Brookings Institution, 1981) p. 698. Prices cited are in current dollars.
9. Calculated from data in U.S. Department of Commerce, *Materials Survey,* pp. 10–11.
10. Peck, *Competition in the Aluminum Industry,* p. 158.
11. Estimated from data on new plants and plant expansion in U.S. Department of Commerce, *Materials Survey,* chapter 3; Langton, *Investment Patterns,* chapters 3 and 4; U.S. Bureau of Mines, "Aluminum Plants Worldwide" (mimeo, Washington, D.C., 1983); and the Aluminum Association, *Aluminum Statistical Review, 1979* (Washington, D.C., 1980) pp. 34–35.
12. Langton, *Investment Patterns,* pp. 99–100.
13. Reliable data are lacking on the relative costs of building new plants at various worldwide locations at the time of the second wave of expansion. Studies by consultants reported that in 1974 the costs for increasing capacity at existing plants in the United States were 69 cents (capital and operating costs); in 1972, such costs were 60 cents for Australia. Most of the higher U.S. costs can be accounted for by higher energy prices. (Author's interviews.)
14. Computed from data in Langton, *Investment Patterns,* table A-1.
15. Ibid., p. 49.
16. For further discussion, see Andrew C. Brown, "Alcan Shakes the Aluminum Market," *Fortune* vol. 107, no. 4 (February 21, 1983) pp. 128–130.
17. See U.S. Bureau of Mines, *Aluminum Plants Worldwide, 1982,* and *Metal Statistics, 1984* (New York, Fairchild Publications, 1984).
18. Computed from Goodwin, *Energy Policy,* p. 698.

19. The term *at risk* comes from Resource Strategies, Inc., "Electric Power Rates for the Aluminum Industry in the Pacific Northwest," a 1984 mimeographed report prepared for the Bonneville Power Administration (see pp. 44–56). This report identifies five smelters in the Northwest with power costs so high that they might be considered to be "at risk." The report also names three other smelters in the TVA areas that are reported to have variable costs comparable to the Northwest plants; all three were operating at 50 percent capacity or less on January 1, 1985.
20. U.S. Bureau of Mines, *Metal Statistics, 1984,* p. 24.
21. "The Direct Service Industries and Their Role in the Regional Economy, A Report by Public Utility District No. 1 of Snohomish County" (September 25, 1985) p. 12.
22. Ibid., p. 13.
23. For an extensive discussion of ad hoc government actions involving these two industries, see Robert Crandall, *The U.S. Steel Industry in Recurrent Crisis: Policy Options in a Competitive World* (Washington, D.C., The Brookings Institution, 1981) chapters 4, 6, and 8; and Robert Crandall, "Import Quotas and the Automobile Industry: The Costs of Protectionism," *The Brookings Review* vol. 2, no. 4 (Summer 1984) pp. 8–16.
24. Bonneville Power Administration, "Selling Power: BPA's Direct Service Industries: Changing Conditions—Changing Needs" (February 1985) p. 3.
25. Ibid., p. 16.
26. Ibid., p. 4.
27. Ibid., p. 5.
28. The events of the late 1970s that led to the passage of the Northwest Power Act of 1980 are found in U.S. General Accounting Office, *Impact and Implications of the Pacific Northwest Power Bill* (Washington, D.C., 1983). See the letter and attachments submitted by Eric Redman on behalf of Direct Service Industrial Customers submitted to the Subcommittee on Energy and Power of the Committee on Interstate and Foreign Commerce, U.S. House of Representatives, 46 Cong. 1 sess. (July 30 and October 9, 1979), as well as the following congressional bills, hearings, and reports: Senate Hearings on S311-41 and S311-42, Committee Publication S311-16, and Senate Report 96-272 (all 1979); and House of Representatives Hearings on H441-31 to H441-35 (1978), H501-70 (1979), and H501-27 and Report 36-976 (1980).
29. For a description of WPPSS, see U.S. Department of Commerce, *Energy and the Primary Aluminum Industry* (Washington, D.C., November 1984) pp. 22–24.
30. Roberta C. Yatie, "6 Aluminum Companies Accept BPA's Incentive

Power, 2 Firms Adjust Output," *American Metal Market* vol. 92, no. 170, p. 1.

31. U.S. Department of Commerce, *Energy and the Primary Aluminum Industry,* p. 27. In the summer of 1983, WPPSS defaulted on its $2.5 billion bond issue—the largest default ever of tax-free municipal bonds. (*Wall Street Journal,* July 10, 1986, p. 21.)
32. *Wall Street Journal,* p. 24.
33. Resource Strategies, Inc., "Electric Power Rates," pp. 9 and 14.
34. The Resource Strategies, Inc., report states that the closure of five smelters would reduce the BPA's power load requirements by 886 megawatts (MW) (ibid., p. 13).
35. This range is derived from the high, median, and low forecasts presented in BPA, "Selling Power," p. 12.
36. These options are discussed briefly but clearly in Bonneville Power Administration, "A Summing Up: BPA Announces Its Conclusions on the DSI Options Study" (June 1985). A more technical discussion is in Bonneville Power Administration, "DSI Options Study: Final Report," parts 1 and 2 (June 1985). This study simulates outcomes for various options under a wide range of scenarios.
37. BPA, "Selling Power," p. 14.
38. Bonneville Power Administration, "Industrial Incentive Rate: Initial Feasibility Study" (mimeo, July 19, 1985).
39. *Metal Statistics, 1985,* p. 24.
40. Peter Zipf, "Aluminum Companies' Reaction Mixed to BPA Offer," *American Metal Market* (May 21, 1985) p. 2.
41. *American Metal Market* (July 23, 1986) p. 2.
42. Organisation for Economic Co-operation and Development (OECD), *Aluminum Industry: Energy Aspects of Structural Change* (Paris, OECD, 1983) p. 33.
43. U.S. Department of Commerce, *Energy and the Primary Aluminum Industry,* p. 35.
44. *Metal Statistics, 1984,* p. 24.
45. OECD, *Structural Change,* pp. 35–36.
46. U.S. Department of Commerce, *Energy and the Primary Aluminum Industry,* p. 36.
47. Ibid., pp. 35–36.
48. Ibid., pp. 36–37.
49. Ibid., p. 38.
50. Ibid., p. 39.
51. *Business Week* (February 4, 1985) p. 36.
52. U.S. Department of Labor, Bureau of Labor Statistics, *Supplement to*

Employment and Earnings Data (July, 1984) pp. 51, 56, and 126. Aluminum employment is for primary aluminum (standard industrial code [SIC] 3334), automobile employment is for motor vehicles and equipment (SIC 371), and steel employment is for blast furnaces and basic steel products (SIC 331).

53. The data on corporate earnings in these paragraphs are from Datex, *Corporate Database* (New York, 1985).
54. Datex, *Corporate Summaries 1984/85,* pp. 135, 225, 452, 479, 783, and 1151.
55. U.S. Department of Commerce, *Energy and the Primary Aluminum Industry,* pp. 28–29.
56. Even though the stockpile goal was 634,000 tonnes as of July 31, 1984, the actual inventory was less than 2,000 tonnes. Such a small inventory relative to the objective indicates the low priority given to aluminum by the Defense Materials System (ibid., p. 1).

3

WESTERN EUROPE: SUBSIDIZED SURVIVAL

CHRISTIAN KIRCHNER

In the introduction to this volume, Western Europe, together with the United States and Japan, was named as a region that had lost its international competitiveness. It would seem to follow that, in such circumstances, the aluminum industry would be—in the popular parlance—a "loser" industry, with smelters operating at significantly less than capacity and even, in some cases, closing. As of 1983, however, few smelters had been closed, and most were operating at full capacity. Because Western Europe accounts for a fifth of the world's capacity, such continued production has had global ramifications.

Western Europe's seemingly contradictory position is largely explained by the public policies of the several European countries. These policies have offset a high-cost power situation by establishing preferential rates for the electricity used in smelting. In addition, state ownership of aluminum companies in several countries permits the use of public subsidies to cover the losses of smelters. In Western Europe the various governments have acted to maintain high domestic production levels of primary aluminum to avoid dependence on imported aluminum and to prevent unemployment. Moreover, the high value of the dollar between 1980 and 1985 has favored domestic production.

Nevertheless there have been some signs of change in these practices, especially following the second oil shock in 1979. Under the pressure of large losses by the aluminum companies, several govern-

---

Christian Kirchner is professor in the Faculty of Law at the University of Hanover, Federal Republic of Germany.

ments have begun programs to close some of the older, less efficient smelters. A gradual shift away from the policy of maintaining the production of primary aluminum is also reflected in the investments by some of the big Western European aluminum companies in smelters located in non-European countries with low-cost energy. But the aluminum companies have not yet matched such participation in overseas smelting projects with significant reductions of capacity in Europe; rather, the overseas expansion is planned to fill the gap between the increasing demand for aluminum in Western Europe and a stable level of domestic production.

Western Europe, of course, is a region, not a country. Aluminum production varies among the twelve primary aluminum-producing countries (in alphabetical order, Austria, the Federal Republic of Germany, France, Greece, Iceland, Italy, the Netherlands, Norway, Spain, Sweden, Switzerland, and the United Kingdom).[1] The national industries taken together have enough common features to permit an overall analysis, but there are differences among them that warrant separate discussion of the more significant national aluminum industries. These discussions are particularly necessary for understanding public policy in the various nations because despite the emphasis on Western Europe as an economic entity with a common public policy, each country has actually adopted different measures with respect to its aluminum industry. Indeed, national rivalries within Western Europe still abound, and the pursuit of immediate national interest at the expense of regional interest has often characterized public policy.

The overall analysis is facilitated by grouping the twelve countries by two characteristics: whether the industry is domestically or export oriented and whether the industry is or is not a major aluminum producer; that is, annual production is in excess of a quarter of a million tons (all tonnage is in metric tons, or tonnes). The groupings are as follows:

	Domestic	*Export*
Major producers	Federal Republic of Germany France United Kingdom Italy	Norway
Minor producers	Austria Sweden Switzerland Spain	Iceland Greece Netherlands

The five major producing countries accounted for 71 percent of the production of aluminum in Western Europe;[2] accordingly, their policies are emphasized. In addition, the chapter focuses on the aluminum industry after the 1973 oil shock. The responses to that dramatic event, however, cannot be understood without a description of the postwar wave of expansion that occurred from 1950 to 1973. That period of expansion therefore is also treated.

Finally, there are three questions of major significance for this chapter: (1) Should there have been a greater reduction of capacity, and are more reductions likely? (2) How has public policy affected capacity reduction? (3) Should the European nations have created large national aluminum industries in the first place, even given the economic circumstances prevailing before 1973? The following pages propose some answers.

POSTWAR DEVELOPMENT OF THE WESTERN EUROPEAN ALUMINUM INDUSTRY

In 1950 Western Europe relied on imports for 40 percent of its total consumption of primary aluminum because annual production in the region was only 243,000 tons. By 1974, production and consumption of primary aluminum equaled about 3.3 million tons (table 3-1).[3] Despite this expansion, however, four of the five major producing countries remained net importers throughout the postwar years, with Norway the only significant exporter.

During the expansion years the market structure of the European industry assumed its present form. Two of the big six producers headquartered in Europe—Alusuisse and Pechiney—followed a strategy of geographical diversification both within Western Europe and throughout the world. The U.S. big three, together with Alcan, all started smelting operations in Western Europe to substitute local production for their previous exports. In the Federal Republic of Germany, Italy, Austria, and most lately France, publicly owned companies currently are important in the market.[4] The remaining firms are relatively small private producers whose operations are limited to one country.

For four of the five major producing countries, the large expansion of the Western European aluminum smelting capacity was a classic case of import substitution. (Norway, which served nearby export markets, is the single exception.) A factor specific to the aluminum industry, however, was the need to protect domestic fabricators from periodic shortages of primary aluminum at the producers' list

Table 3-1. Production and Consumption of Primary Aluminum in Western Europe, Selected Years, 1950–1983 (thousands of metric tons)

	Domestically oriented								Export oriented		Minor producing countries[a]		Total Europe	
	F.R. Germany		France		United Kingdom		Italy		Norway					
Year	P[b]	C[b]	P	C	P	C	P	C	P	C	P	C	P	C
1950	27.8	49.8	60.6	55.2	29.9	183.8	37.0	47.5	45.3	9.5	43.2	50.6	243.8	396.4
1956	147.4	172.8	149.8	134.8	28.0	280.6	63.7	71.7	92.7	15.3	115.5	165.9	597.1	841.1
1962	177.8	301.9	294.5	235.6	34.6	285.4	80.9	115.0	208.9	24.1	184.4	259.6	981.1	1,221.6
1968	257.4	539.3	365.7	293.5	38.2	388.4	142.3	217.0	462.8	54.0	431.0	516.9	1,697.4	2,009.1
1974	688.9	872.5	393.3	480.0	293.1	492.1	212.3	375.0	648.2	93.5	914.1	960.1	3,149.9	3,273.2
1980	730.7	1042.3	431.9	595.9	374.4	409.3	271.2	458.0	662.6	118.4	1,127.4	1,084.8	3,598.2	3,708.7
1983	743.3	1085.0	360.8	613.4	252.5	323.4	195.7	430.0	715.4	128.6	1,059.4	730.7	3,327.1	3,311.1

Sources: Metallgesellschaft, *Metallstatistik* (Frankfurt/Main) various issues; also the English version, *Metal Statistics.*

[a] Austria, Greece, Iceland, the Netherlands, Spain, Sweden, and Switzerland.

[b] P = production; C = consumption

price. Indeed, domestic production in the various countries was intended to provide a secure supply of primary aluminum to the nations' fabricators.[5]

In common with other regions of the world, the expansion of the domestically oriented aluminum industry in Western Europe was based on the availability of low-cost power. Prior to the oil shocks, the Western European price of electricity for aluminum smelters was not much above the price elsewhere. In addition, long-term power contracts, based on optimistic expectations and thus providing lower rates, were also available. Electricity charges thus were a small fraction of total costs.[6] Additionally, there were transportation cost savings in locating the smelters close to the industrial centers of Western Europe.[7] Paradoxically, it appears that aluminum smelting was promoted precisely because it was energy intensive and so could consume large blocks of electricity immediately, thus supporting the construction of nuclear power plants. Domestic costs to construct a smelter were further reduced by the low exchange rates of some European currencies vis-à-vis the U.S. dollar, although that advantage was partially offset by the higher costs (in domestic currency) of imported raw materials. Tariff considerations played a minor role in the location decisions.[8]

The expansion of Europe's export-oriented aluminum industries was also based on low-priced electricity. Norway, Iceland, and Greece all had available hydropower; in the Netherlands, electricity was obtained relatively cheaply from natural gas-fired and nuclear power plants. In Greece, the availability of domestic bauxite was also an important factor in its decision to build an export-oriented smelter.

The expansion of smelting capacity in Western Europe came to a sharp halt in 1975 when the impact of the first oil price shock was fully realized. (A few smelters had been expanded after 1974, and one new smelter—planned in the early 1970s—went into production.) Since 1975 the European industry has been characterized by relatively stable levels of production; modest modernization of old smelters; large price increases for electricity, which have been partly mitigated by long-term power contracts; and rising costs for environmental protection. Yet by 1984, Western Europe still had production that appoximately equaled consumption. The ways in which each of the five major producer countries and the seven minor producing countries handled the 1930–1975 expansion and their responses to the 1973 and 1979 oil shocks are discussed in the sections that follow.

The Federal Republic of Germany

The prewar aluminum industry in Germany was partially destroyed by World War II bombings, and Allied restrictions prevented reconstruction until 1950 when the four prewar smelters in the Federal Republic of Germany were rebuilt.[9] By 1953 production had surpassed prewar levels. Vereinigte Aluminium-Werke-Aktiengesellschaft (VAW), a state company formed in 1917 to organize World War I production, owned three of the smelters; Alusuisse owned the fourth.[10]

The next step in Germany's postwar expansion came in 1957 with the VAW acquisition of large brown coal reserves for thermal power generation to supply its two existing smelters and to permit the construction of a new smelter. (Brown coal or lignite is a grade of coal between peat and bituminous coal that is used primarily for boiler fuel. It has a low thermal content and therefore a low price relative to bituminous or black coal.) VAW also concluded a long-term power contract for three of its four smelters with Rheinisch Westfaelische Elektrizitactswerke AG (RWE), a large supplier of power. These four smelters together with the one owned by Alusuisse were repeatedly modernized and expanded. As a result, by 1969 annual production capacity had reached 269,000 tons.

The expansion of capacity, however, still lagged well behind the growth in demand; in 1962, for example, imports, largely from Norway and North America, accounted for approximately one-half of the primary aluminum consumed in Germany. Such a great dependence on imports was seen as a potential threat to German industrial development, particularly because there were recurring shortages of imported aluminum. Yet the unavailability of low-cost electricity limited expansion—that is, until the late 1960s, when black coal and nuclear energy became available for low-cost power generation and additional smelters were planned to make use of them.

Because of competition from low-priced oil, coal mining in the Rhein-Ruhr region of Germany by this time had become a depressed industry. Coal mining companies were seeking additional outlets for their black coal; power generation for aluminum smelters promised to be an important growing market. (Note that although smelting is not labor intensive, this supplying industry is labor intensive.) Concurrently, nuclear power technology had progressed enough that German experts were confidently predicting that German industry could be supplied with internationally competitive power from nuclear generation. Thus, two considerations—markets for the ailing German coal mining industry and the expectation of cheap nuclear power—combined to promote a large expansion of aluminum smelting capacity in the late 1960s and early 1970s. Five new smelters were constructed, and plans were made for the further expansion of existing smelters.

The five new smelters were owned by four new entrants into the primary aluminum industry in Germany: Kaiser and Reynolds, in a joint venture, and three German companies.[11] Three of the five smelters were planned to use power from proposed nuclear facilities, and two were to use power from black coal generation. The black coal-generated power was provided by RWE, whose shareholders were the municipalities of the Rhine-Ruhr region, which would benefit from using black coal. The City of Hamburg also became involved in the aluminum industry as part owner of two of the smelters, which it constructed to aid local economic development. All of the smelters were given long-term contracts for electric power at low rates.

By 1969 the planned expansion of capacity, either through increases in capacity at existing smelters or through the construction of new smelters, was quite striking. If all the plans had been carried out, German smelter capacity would have expanded from 285,000 tons in 1969 to 786,000 tons in 1973. As a result of a number of political decisions and uncertain economic assumptions, these ambitious plans were revised downward several times between 1969 and 1973. Nevertheless, smelting capacity in the Federal Republic of Germany had reached 745,000 tons by the end of 1973.[12]

In 1975, when the full effects of the 1974 oil shock had become apparent, all expansion plans were delayed indefinitely. In that year, consumption of primary aluminum fell by 19.3 percent although production declined only 1.6 percent. German producers maintained their preshock output levels despite large losses from the rising prices of imported alumina and bauxite. In other circumstances, such losses might have been recouped by charging higher prices; the steady weakening of the U.S. dollar against the German mark in the late 1970s, however, reduced import prices and precluded German producers from passing on higher costs by raising their prices. The price–cost squeeze was further intensified by a decline in demand from German fabricators as a result of the general recession following the 1973 oil crisis.[13] Yet German smelters did not reduce production despite their losses, largely because inflexible long-term contracts for electricity involved heavy penalties for reducing their power consumption. The losses led ultimately to a reorganization of the industry with VAW, Alusuisse, and Alcan buying out the smaller German producers.[14] One smelter was entirely closed, another was partially closed, and construction stopped on a third. The larger question of what to do with the German aluminum industry was also much debated and is discussed later in this chapter.

France

Production of primary aluminum in France increased from 60,600 tons in 1950 to 393,000 tons in 1974. Two French companies—Pechiney and Ugine—together owned the eleven domestic smelters, all but one of which had less than the 100,000-ton capacity common elsewhere. Even though output expanded nearly sevenfold, however, imports still accounted for one-fifth of French consumption of primary aluminum in 1974. In addition, further expansion even before the first oil shock was limited by the high price of incremental electricity. Aluminum producers also were unable to acquire or construct their own generating facilities because the nationalization of all electricity generation plants in France in 1948 reserved electricity production to the national system. In 1950, however, Pechiney took advantage of the natural gas fields discovered in the region of Lacq to construct a 103,000-ton capacity smelter—which was the one large, new French smelter.[15] Expansion of the existing ten smaller smelters accounted for the remaining growth in capacity.

In search of cheap electricity, the two French aluminum companies also constructed smelters in Cameroon (a former French colony), Greece, and the Netherlands.[16] Power for the Netherlands smelter was to be supplied by a nuclear power plant; the two other smelters depended on hydropower. Pechiney also became part owner of a U.S. smelter.

Pechiney and Ugine cooperated closely in a number of ventures, and the two companies merged in 1971.[17] Since then, no other companies have entered the French aluminum industry, partly as a result of a national policy of reserving key industries for national companies. This policy, together with the unavailability of low-priced electricity, discouraged the big six from building smelters in France.

In 1975 expansion came to a halt, as it did elsewhere in Europe, with France's aluminum production capacity at 410,000 tons. Since then, Pechiney has cautiously restructured its domestic smelters by modernization, the application of energy-saving techniques, and the closing of a few small, very old smelters.[18] The net effect has been no change in capacity. Moreover, Pechiney's nationalization in 1981 has not changed its corporate strategy. Unlike the German smelters, Pechiney has been able to operate its domestic smelters without large losses, in part because of the favorable electricity rates that have accompanied the power surpluses of the 1980s in France (see the discussion later in this chapter). In 1984, Pechiney's new president announced plans for a "leaner, more efficient company": four smelters were to be closed, the remaining seven were to be modernized, and consideration would be given to constructing smelters abroad.[19]

The United Kingdom

The United Kingdom was a latecomer in the expansion wave: until 1970, production of primary aluminum was still less than 40,000 tons. (By this time, production in the Federal Republic of Germany and in France was over 350,000 tons.) Of the 410,000 tons of aluminum Britain consumed, imports supplied approximately 90 percent. This huge import gap was the basis of a political decision to encourage the building of smelters by offering favorable electricity rates for the domestic production of aluminum.[20]

Initially, the British incentive succeeded. Three new smelters began production during 1970–1972: one owned by Alcan; another by the British Aluminum Company (BAC), the only previous producer; and a third owned by a joint venture of Kaiser and two British companies. But technical difficulties, strikes at the smelters, and an inadequate supply of energy (partly traceable to a miners' strike) prevented the new smelters from operating at full capacity between 1971 and 1974. By 1975, however, Britain's dependence on imported aluminum was down to 25 percent, and thus the goal of reducing dependence on imports was realized.

Yet achieving that goal turned out to be at the cost of large losses for BAC. The biggest source of these losses was the large cost overruns incurred in the construction of a nuclear power plant being built by the North of Scotland Hydro Electricity Board to serve the new BAC smelter. The contractual arrangements between the board and BAC allowed part of these increased costs to be passed on to BAC; as a result, BAC eventually closed its new Scottish smelters.[21] Subsequently, BAC was taken over by Alcan.[22]

As of 1983 two small, older smelters with a combined capacity of 39,000 tons were continuing to operate despite high costs. The other two operating smelters, built in the early 1970s, are larger and more modern and have lower costs. All four smelters, however, depend on special electricity rates for their continued operation.

Italy

In the early 1960s, the Italian aluminum industry consisted of three companies—Mican, Alusuisse, and Montecatini—operating six small smelters. Domestic production amounted to 80,000 tons and supplied about two-thirds of domestic consumption.

As in other European countries, import dependence was a key factor in initiating an expansion of Italy's domestic capacity. To reduce such dependence, construction of a new, 125,000-ton smelter to be operated by Alsar, a joint venture of the state and Montecatini, began in 1965.[23] The plans for this smelter were based on the use of nuclear energy that was predicted to be relatively low in cost. Construction also began on a second smelter with only 39,000 tons of capacity.[24] Finally, plans were drawn up for two other smelters as part of an expansion of national capacity from 80,000 tons to as much as 562,000 tons.

The small smelter began operation in 1970, and the Alsar smelter began operation in 1973. Although there was also expansion of the capacity of existing smelters between 1970 and 1974, plans for the two additional smelters were abandoned in 1974.[25] Italian capacity reached about 450,000 tons by 1980 with Alsar operating at only 53 percent of capacity at this time. Imports continued to be significant, supplying about 40 percent of domestic consumption.

The low rate of use of domestic production reflected the high cost of such production relative to the international price of aluminum—high costs that were mainly due to increased electricity rates.[26] About 80 percent of Italy's electricity was oil generated; thus, the two oil shocks had dramatic impacts on the cost of electricity generation in that country.

Only one smelter was closed, however—an old one owned by Alcan. In 1973, MCS, a subsidiary of the state holding company, EFIM, took over all the Montecatini smelters.[27] The industry thus became fully nationalized, apart from the Alusuisse joint interest.

As the losses increased, the parties involved discussed closing more smelters, but the restructuring of the Italian aluminum industry turned out to be politically difficult.[28] The reorganization plans, which were never implemented, allowed Alusuisse to participate in another newer smelter to offset the closing of its jointly owned smelter. MCS, however, with large losses and sizable government subsidies, was a less than ideal partner for a new joint venture.[29] And in any case, there was substantial political opposition to closing the smelters, despite the large losses that were being covered by Italian taxpayers.

Norway

Norway has the largest export-oriented industry in Western Europe. After 1950 the country resumed its prewar growth in aluminum production, increasing from 45,000 tons in 1950 to 200,000 tons in 1960. Eighty percent of the output was exported, mainly to the Federal Republic of Germany, Denmark, the United Kingdom, and Italy. Three companies were active in aluminum smelting: Ardal og Sunndal Verk A/S (ASV), a state company; Det Norske Nitridaktieselskap (DNN), a joint venture of Alcan, British Aluminum, and Pechiney; and Norsk Aluminium Co. A/S (Naco), a joint venture of Alcan and various domestic owners.[30]

Aluminum smelting in Norway, which has always been based on cheap hydropower, was and is largely under the control of the international aluminum companies. These companies supplied the Norwegian smelters with alumina and the expertise to construct and operate smelters. Electricity for the smelters was supplied either by generating plants owned by the aluminum companies or by NVE, a state company. This pattern continued when Alusuisse constructed a smelter in 1962 and when Alnor A/S, a joint venture of Norsk Hydro and Harvey Aluminum, opened its smelter in 1967.[31]

By 1972 the country's smelting capacity had reached 700,000 tons, making Norway and Germany the only European countries with capacities of more than half a million tons. This capacity was not fully used, however, because of the difficulties in selling 700,000 tons of primary aluminum on the European market. Plans for further expansion were postponed, and one old smelter was closed in 1973.

Yet beginning in 1974 there has been extensive debate in Norway about whether to expand aluminum capacity still further. Parliament has played a large role because it must approve the construction of hydropower capacity, which is necessary for any expansion.[32] The reservations that have been expressed during the debate on expansion are related to the uncertainties of the export market.

Norway is not a member of the European Economic Community (EEC), whose members nevertheless account for most of its exports. In addition, Norway negotiates annually with the EEC the amount of Norwegian exports that will be allowed by the EEC on a tariff-free basis.[33] Given this vulnerability, the Norwegian Parliament delayed the expansion of smelting capacity.[34]

In 1984 the three aluminum producers then operating in Norway (ASV, DNN, and Naco) announced plans to expand capacity by 105,000 tons from their current 620,000 tons. Some of the expansion would be achieved by plant improvements to reduce the kilowatts required per ton; the remaining additions to capacity would require more electricity. These plans have been under study by the government since that time.[35]

The Minor Producing Countries

Three countries—the Netherlands, Iceland, and Greece—created small, export-oriented aluminum industries serving European markets. In each case, although the particular circumstances varied among the three countries, the new smelters that were constructed reflected a perception that cheap electricity would enable a national aluminum industry to be internationally competitive.

The Iceland story is the simplest. The country had considerable hydropower capacity, and in 1966 it offered Alusuisse low electricity rates for fifteen years to encourage the company to build a smelter there. Alusuisse built a relatively small smelter—with a 66,000-ton capacity—and Alusuisse took all the production from the smelter to serve its customers.[36] When the fifteen-year initial power contract expired in the early 1980s, the Icelandic government sought to raise its electricity price to the smelter significantly (the smelter was the only big industrial user of electricity in the country).[37] Alusuisse resisted the increase and threatened to close the smelter, setting off considerable political controversy. A new compromise rate was finally negotiated, but it is not surprising that the plans of Norsk Hydro, a Norwegian power supplier, to build a second smelter were abandoned.

Greece also has a one-smelter aluminum industry, although it is one of the only Western European countries with bauxite deposits and exports both bauxite and alumina. Aluminum de Greece, a joint venture of Pechiney and local investors, began operating in 1966 using local alumina and electricity from a new hydropower facility constructed by the Greek government.[38] The smelter was expanded after 1970, and the industry has continued to export aluminum in sufficient volume to utilize fully the smelter's capacity.

In the Netherlands, two smelters were built in 1966 and 1971. The first, initially a joint venture of two Dutch companies and Alusuisse, obtained its electricity at low rates from a generating plant that used natural gas. (Alusuisse subsequently sold its share to its Dutch partners.) The second smelter, built by Pechiney, was based on the prospect of cheap nuclear power.[39] Initially, the production from these two smelters served the Dutch market, but by 1974 the Netherlands was a net exporter, largely to the Federal Republic of Germany. Since 1974 the expansion has stopped, and the industry has continued to use its capacity fully by selling in the export market.

Until the early 1970s, Spain also had a small primary aluminum industry serving domestic markets. A new smelter, jointly owned by INI, a state company, and Alcan and Pechiney, began operation in 1979, coincidentally with a downturn in demand in export markets.[40] When the smelter was planned, it was assumed that cheap power would be available from brown coal thermal generation; in fact, that power turned out to be quite expensive because of the underestimation of both construction and operating costs. The new smelter had large losses, and the combination of state and private ownership led to disputes as to how the losses were to be divided among the partners.[41] Since the late 1970s, other Spanish smelters have also accumulated sizable losses.

Austria, Sweden, and Switzerland are perhaps the most fortunate among the twelve European countries with primary aluminum industries. Their industries were all small, prewar legacies with low-cost hydropower that were rarely expanded because there was no possibility of additional hydropower. They were also spared the big country pretensions that led to extensive construction of nuclear power. Subsequently, the oil shocks of 1973 and 1979 have had less impact on the industries in these three countries.

THE INTERNATIONAL COMPETITIVENESS OF THE EUROPEAN ALUMINUM INDUSTRY

Although the exact timing varied for each particular country, the European aluminum industry had two distinct phases: (1) an expansion wave up to 1975 and (2) the maintenance of output despite losses after 1975. What is examined in this section is whether either the expansion or the continued operations that followed made economic sense in the four countries with large domestically oriented aluminum industries: the Federal Republic of Germany, France, the United Kingdom, and Italy.

Before the Oil Shocks

A common explanation of the aluminum industry's expansion has been the character of exchange rates in the 1960s. Until 1972 the United States and other countries operated in a regime of fixed exchange. In the 1960s and early 1970s these fixed exchange rates were held to have overvalued the dollar. This factor is of considerable importance because international aluminum prices were quoted in dollars. The effect of an overvalued dollar was to raise the prices of imported aluminum relative to domestic production and thus encourage the expansion of domestic production. To be sure, an overvalued dollar also raised the price of imported bauxite and alumina. Yet, as Goto shows in chapter 4, the net effect of an overvalued dollar (or, in his case, an undervalued yen—the other side of the overvalued dollar) was to raise the prices of imported aluminum more than the prices of the imported materials necessary to produce it and so favor domestic production.

In attempting to understand the Western European expansion, particular stress should be placed on the capital costs of constructing a smelter. These were expenditures incurred in local currencies, which were undervalued. The big three had the choice of expanding in Europe or expanding in the United States. Europe's lower construction costs may have been particularly attractive to these producers because hydropower was increasingly hard to find in the United States and infrastructure investments were required to make use of the cheaper power of Australia and Brazil; Canada, on the other hand, seemed the preserve of Alcan. Joint ventures with European partners were common, reducing both the risk and capital requirements.

Yet along with an overvalued currency, Europe offered undervalued electricity, primarily on the promise of cheap nuclear power. The stress is on the promise; at the time the aluminum industry's expansion occurred, nuclear power plants were either under construction or in the planning stage. In the end the estimates of construction costs turned out to be highly optimistic, and environmental and safety costs were generally underestimated.

An overvalued dollar and undervalued power tended to be incorporated into the formal analysis of the economics of primary aluminum expansion. But more intangible considerations also led Western European governments to promote the expansion of the aluminum industry. Consumption of aluminum grew rapidly throughout Europe in the 1960s. If aluminum could be produced domestically, there promised to be four benefits to the national economy. First, there would be significant balance of payments savings. The United Kingdom and Italy, and, to a lesser extent, France, had periodic balance of payments crises. Although the Federal Republic of Germany did not, even in that nation some value was attached to balance of payments savings. Import substitution was considered to be easier than making equivalent export gains. Second, the rapidly growing domestic fabricators might be helped by domestic primary aluminum production. Because primary aluminum is a standard commodity, the fabricators' gains were the intangible ones of service, delivery times, and priority, should such aluminum be in short supply. There was also a fear of too great a dependence on imports, although producing aluminum domestically only pushed the dependence one step back to imported alumina and bauxite. Third, creating a domestic aluminum industry would create jobs, although aluminum smelting is not very labor intensive and there may have been other measures that would have done more for employment. Finally, smelters were often located in isolated regions with particular problems of lagging development and so could serve as instruments for regional development.

No one of these benefits seems to have been decisive in the building of any particular smelter. Yet their combined effects, together with the overvalued dollar and undervalued power, explain the postwar expansion. As noted earlier, particular stress should be placed on the relatively inexpensive power that was available and the related importance of aluminum smelters as large sources of demand for the output of the nuclear power plants. Still another factor was that governments could directly or indirectly promote the expansion of aluminum capacity because they controlled the construction of power plants. In Italy and Germany the government impact on the industry was strengthened even further through the control of state-owned companies—MCS and VAW. And in Germany the Rhine-Ruhr municipalities, through their ownership of the South German power grid, and the city of Hamburg, in its joint ventures, played important roles in promoting the expansion of aluminum capacity regionally.

Governments in general have difficulty finding appropriate instruments to promote an industry in a market system. Tax incentives and subsidies are often clumsy tools, and the use of trade protection runs counter to the prevailing trends to establish a common market and lower trade barriers. In the late 1960s and early 1970s, a publicly owned power system and state ownership of smelters appeared to be direct instruments that could be used to promote the aluminum industry—another proof that the role of convenience in public policy should never be underestimated.

The European aluminum industry probably was not internationally competitive at the margin before the oil shocks. To verify this assertion is complicated because it involves assessing the proper discounts for risk to be applied to the undervalued electricity and overvalued dollar of the period, as well as the value of the intangible benefits of expansion of an aluminum industry. It may be said in hindsight that the risks were underestimated. Even at the time, however, the expansion wave seemed to have been based on a very optimistic reading of the future. Particularly puzzling was the United Kingdom expansion that came after 1974 when it was clear that nuclear power was not fulfilling its promise.

The stress should be on bad decisions at the margin, however. There was a fourfold expansion of capacity from 1960 to 1974—about 2.4 million tons. It seems unlikely that Australia, Brazil, and Canada, together with the other countries with low-cost electricity, could have accommodated such a high rate of expansion without encountering sharply rising construction costs for power plants and infrastructure. The Australian experience, described in chapter 5, shows that "boom" periods tend to impose high costs. Some expansion in Europe would have been economically efficient; there was simply too much capacity created at too fast a rate.

Italy and the United Kingdom accounted for too much of the expansion. The Federal Republic of Germany may be another case of excessive expansion; its aluminum industry came to equal Norway's in size—and Norway was the premier low-cost energy country of Europe. Among the smaller countries, Spain managed to establish an export-oriented industry just when the demand for aluminum was falling and the price of electricity was rising.

Predicting the state of the European aluminum industry had the oil shocks never occurred is an interesting puzzle. The false promise of cheap nuclear power would still have been revealed—an important factor, as the preceding accounts indicate, in the United Kingdom, Italy, and the Federal Republic of Germany. And the overvalued dollar would have disappeared, making import competition more important. Still, the losses would have been more modest, and, with special electricity rates and state ownership in the Federal Republic of Germany, France, and Italy, they might never have become apparent.

After the Oil Shocks

Because the oil shocks did occur, the prices the smelters paid for electricity rose sharply. Table 3-2 shows the electricity rates aluminum smelters were charged in various countries in 1977 after the first oil shock and in 1982 after the second, demonstrating the sharp rise in electricity prices. These rates do not reflect the marginal

Table 3-2. Electricity Rates in Current U.S. Dollars for Aluminum Smelters, 1977 and 1982 (mills per kilowatt-hour)

Country	1977	1982
Major producers		
Domestic		
Federal Republic of Germany	8–16	24.5 (medium)
		12.5 (lowest)
France	9–15	26 (medium)[a]
Italy	20	35 (medium)
		20 (lowest)
		50 (highest)
Export		
Norway	5	9.5 (medium)
		3 (lowest)
		20 (highest)
Minor producers		
Domestic		
Spain		22.4 (medium)
		55 (highest)
Export		
Netherlands		12–16
Iceland		6.5
Greece		20 (medium)

Sources: Hideo Hashimoto, *Bauxite Processing in Developing Countries* (London, Commonwealth Secretariat, 1983); The World Bank, *Case Studies on Industrial Processing of Primary Products,* vol. 1 (Washington, D.C.) p. 43.

[a] Stated in current U.S. dollars, the rate for France in 1982 does not take into account the rate reduction granted by the French power system EDF to Pechiney in 1983, which resulted in a rate of about 15 mills/kWh. See "Laenderkurznachrichten. Frankreich" (Country News. France), *Aluminium,* vol. 59, no. 8 (1983) p. 562. Data for the United Kingdom, Austria, Sweden, and Switzerland are unavailable.

social cost of electricity. European countries tended to follow the policy of averaging low-cost and high-cost power in setting rates and giving special rates to aluminum smelters. But, of course, other countries also followed a policy of cost averaging and special rates.

If actual electricity prices are considered, Western Europe is generally in the range reported in chapter 1. That means that existing plants on average are likely to be able to continue to operate and to cover their operating costs. The prices paid for electricity in Western Europe are well below the electricity rates, for example, in Japan, rates that led to the massive capacity reduction described in the next chapter.

Still, in Western Europe, about 8 percent of the power used in primary aluminum production comes from oil generation.[42] An OECD study puts the cost of oil-generated electric power at 66 mills/kWh compared to 48 mills for coal, 39 mills for nuclear power, and 24 mills for hydropower (all in 1981 dollars).[43] On this score alone, the 8 percent of oil-generated power must be considered to be high-cost power. The proportion of oil-generated power varies among countries (table 3-3). In addition, other fuel sources may be important for particular countries. For example, the dependence on coal by the Federal Republic of Germany is significant; nuclear power ranks second, but that ranking understates its importance for the aluminum industry—many of the expansion plans for smelting assumed nuclear power capacity whose construction has been delayed or abandoned. France is unique in the use of natural gas, but its uniqueness is limited to one 103,000-ton smelter at Nogueres where gas discoveries led to electricity generation for aluminum smelting.

Table 3-3. Distribution of Sources of Electric Power for Aluminum Smelting
(percentage)

Country	Hydro	Natural gas	Coal	Nuclear	Oil
Federal Republic of Germany	11	—	67	16	3
France	25	37	11	4	23
United Kingdom	11	—	35	54	—
Italy	20	—	—	—	80

Source: Organisation for Economic Co-operation and Development, *Aluminum Industry: Energy Aspects of Structural Change* (Paris, OECD, 1983) pp. 35–37. Given long-term power contracts, it is possible to associate smelters with particular sources, although given the prevalence of power exchange (power coming from different sources), any assignment has an arbitrary character.

Again, for France, nuclear power is more important than its current percentage indicates; it was the construction of nuclear power capacity that created the power surplus that lowered electricity rates to smelters, even though the nuclear power plants did not directly serve the smelters. The United Kingdom's high proportion of nuclear power reflects the new smelter in Scotland that has since been closed.

Another feature of the European industry is the number of small smelters. Smelters with capacities of less than 50,000 tons constitute 17.3 percent of European production.[44] In contrast, the United States and Canada have only 1.6 percent of their capacity in such small smelters, which typically have operating costs that are 10 to 20 percent higher than those of larger smelters. An annual capacity of 100,000 tons per smelter has been considered the minimum efficient scale for some two decades. The combined effect of small-capacity smelters and oil-generated electricity probably means that approximately one-fifth of the European capacity can be considered high-cost capacity. Current operating costs for those smelters could not possibly be covered in an unsubsidized world of free international trade.

There are several additional complications related to power costs in Western Europe. First, as described in the note to table 3-2, in 1982 the French power system EDF offered Pechiney a special reduced rate of 15 mills/kWh. This reduced rate was offered because France had excess power capacity as a result of its very large expansion of nuclear power.[45] The surplus power could have been sold abroad to Italy and the Federal Republic of Germany at much more than 15 mills. Such international sales of power are technically feasible; France already sells some power to the Federal Republic of Germany and to Switzerland. The barrier to such sales appears to be created by domestic power companies that have resisted foreign competition in their long-protected markets.[46]

The Federal Republic of Germany represents another anomaly in the European power market. The smelters that were built in that nation have provided a market for the declining black coal industry. In addition, German power companies are obligated to use a high percentage of domestic black coal in their thermal plants; such coal is more expensive than imported coal.[47] That requirement raises power costs in Germany to all users, including aluminum smelters. But although the aluminum smelters are subsidizing the German coal mining industry, they in turn have benefited from the long-term power contracts signed prior to 1975. To be sure, rates to smelters have gone up, but the extent of the rise has been limited by the long-term contracts. The power companies sued unsuccessfully in the German courts to set aside the contracts after the second oil shock.[48] The increases they did obtain came largely from out-of-court settlements, and they were much smaller than the utilities sought. Eventually, these long-term contracts will expire; the case of the Ludwigshafen smelter, whose long-term contract expired in 1982, may be instructive in predicting what will happen in such instances. The continued operation of the smelter was threatened, but public subsidies and a new favorable long-term power contract served to prevent its closure.[49] This example suggests that smelters will be kept in operation at the expense of German taxpayers and consumers.

The power situation in the Federal Republic of Germany is further complicated by a shortage of base load capacity, which is the exact opposite of the situation in France. To meet that shortage, the German government favors an expansion of nuclear capacity. The cost of constructing nuclear plants has risen substantially, however, in part because of stricter environmental protection laws. This situation makes special electricity rates for aluminum smelting very costly to the economy.

The Ludwigshafen smelter is an example of the tendency throughout Western Europe to prevent the closure of smelters through government aid. The subsidy mechanism varies from the direct absorbtion of losses (Italy) to special electricity rates (the United Kingdom). Throughout most of Europe, the marginal social cost of electricity is very high; expansion of capacity is largely accomplished by building nuclear plants. In high-cost power countries such as the Federal Republic of Germany, Italy, and the United Kingdom, a smaller aluminum industry would spare these countries the social cost of building more power capacity as the demand for electricity from other consumers grows. France, although currently enjoying a surplus of power, is only a temporary exception. The overbuilding of nuclear power capacity there resulted in a low marginal cost of electricity; economic growth, however, should eventually increase the demand for power and so eliminate the current surplus that justifies low rates to smelters.

Yet the tendency throughout Europe is to avoid capacity reduction and operate existing plants at full capacity. This policy is quite apparent in the behavior of aluminum production output from 1980 to 1983 during a decline in demand (table 3-4). Comparing Europe as a whole to the United States is particularly instructive because

Table 3-4. Changes in Production, Consumption, and Gross National Product, 1980–1983

Region	Percentage change in production[a]	Percentage change in consumption	Percentage change in GNP
Europe	−6.2	−9.4	−0.1
United States	−29.6	−27.3	−1.7
Japan	−67.9	−12.9	0
World	−15.9	−14.1	0

Source: Resource Strategies, Inc., "Electric Power Rates for the Aluminum Industry in the Pacific Northwest" (Philadelphia, November 1984) pp. 33 and 36.
[a] 1980–1982.

the two regions have roughly comparable operating costs. Yet the decline in output was only 6.2 percent in Europe and 29.6 percent in the United States. To be sure, Europe has had a smaller decline in its real gross national product (GNP) (0.1 percent to the U.S. 1.7 percent) and a smaller decline in aluminum consumption (9.3 percent to the U.S. 27.3 percent) than the United States. But with free trade, high-cost regions should show a decline in output. Japan followed this pattern; Europe did not. Indeed, a consultant's report concludes, "It seems fairly obvious that the finger is pointed decisively at Western Europe, which made only a modest cutback in production in both absolute and percentage terms, although many of its plants are fundamentally high cost when stripped of various government support measures."[50]

The differential in output response between the United States and Europe is sometimes attributed to the behavior of the dollar, which in the early 1980s was commonly regarded as overvalued. An overvalued dollar favors production in Western Europe. But the dollar can be only a part of the explanation of European high output because the aluminum industry still had losses or benefited from special electricity rates.

This analysis is focused on the major domestically oriented industries of Europe. The export-oriented aluminum industries of Norway, Iceland, and Greece have remained internationally competitive in their present size, and expansion has largely stopped. This is a rational response to the rise in incremental costs for additional power and to the actual and potential competition of the ABC countries.

Spain and the Netherlands are the exception; both attempted to create an export-oriented aluminum industry without hydropower.

Not surprisingly, both ran into economic problems, which were most acute in Spain.

PUBLIC POLICY

Public policy created the European aluminum industry; public policy maintains it. Although in many respects valid, those conclusions are probably too sharply drawn to apply to a diverse industry spread among twelve countries. A case could be made that, even if governments had not intervened, a significant fraction of the capacity would have been built in the expansion wave of the 1960s and 1970s anyway and these smelters would have maintained output in the 1980s. But it was government promotion that made the Western European aluminum industry so large and enabled it to keep operating at almost full capacity. The government role was particularly decisive in the United Kingdom, Italy, Spain, the Netherlands, and the Federal Republic of Germany. Norway, Iceland, and Greece—all countries with supplies of available hydropower—represented internationally competitive locations; these industries did not need government promotion. France is more difficult to categorize, but even here the overcapacity of the government power system has been an important factor in the development of the industry.

The reasons why the aluminum industry was selected for special promotion have already been discussed, but they bear repeating. Aluminum was not the subject of an explicit policy of industry promotion, nor were there debates on the advantages of promoting aluminum relative to other industries. Rather, government policy interests were closely related to the promise of nuclear power, particularly in the Federal Republic of Germany, the United Kingdom, and France. Nuclear power, it was thought, would provide an unlimited supply of cheap power to Europe, and thus the aluminum industry would be internationally competitive. But why not wait until the promise was fulfilled to expand a domestic aluminum industry? Because nuclear power plants are economically built on a large scale, expanding this source of power could lead to overcapacity, as in fact happened in France. Yet common wisdom held that if new aluminum smelters were in place when the nuclear capacity began operating, they would immediately take large blocks of power and thus provide a demand for the nuclear power plants. There would be no awkward period of politically embarrassing overcapacity. In the Federal Republic of Germany there was an additional rationale for expansion: providing a market for thermal power generated from black coal, an important high-cost industry in economic trouble. The rapidly growing demand for aluminum by domestic fabricators appeared to guarantee the sale of the output of new smelters. The gain in the balance of payments by lowered imports was considered another benefit, particularly in the United Kingdom. And the added employment in smelters, although small, was always welcome. But the decisive factor for singling out aluminum was related to the promise that nuclear energy would provide low-cost power.

The oil shocks and the sharply rising cost of nuclear power plants made the expansion wave and its rationale a serious economic mistake. European countries were suddenly facing sharply rising costs for incremental power that had to come largely from oil generation or high-cost nuclear plants. Today, the mistakes that have already been made are "sunk" costs—they can be disregarded. But many European smelters still do not cover operating costs. Why are they not closed?

Part of the answer lies in the way the losses are absorbed. To the extent that uneconomic electricity rates are subsidized, the costs are lost in the accounts of the power authorities. And where the losses are in state-owned companies, as in Italy, they are rationalized as merely temporary losses attributable to economic downturns. Finally, smelters are often the major employers in areas away from major industrial cities, and local politicians and unions resist any closings.

Western European public policies thus determine in large part the fate of the aluminum industry in that region. The Federal Republic of Germany is perhaps the most complex example because some of its policies—restrictions on the import of power, the environmental protection regulations, the requirement to use German coal—raise power costs, whereas others tend to lower power costs—notably, charging smelters much lower rates than other industrial users. The principal elements enumerated for Germany are also present in the United Kingdom, although the British coal industry has been subsidized more directly. France is more a case of the overbuilding of nuclear power in the past; Italy is a case in which all available power is simply high-cost power.

Maintaining output and avoiding plant closings is a general policy that is carried out in varying degrees in Western Europe. It is most marked in Italy, with its state-owned companies, and in France, where the continued operation of small smelters reflects the objective of maintaining local employment. But throughout Western Europe government aid has been used to maintain employment in industries in economic difficulty. This policy is in sharp contrast to that of Japan, where the emphasis is on shifting labor and capital from industries that are losing their national competitiveness to those that are gaining competitiveness. As the story of the Japanese aluminum industry demonstrates, Japan has been willing to engage in major capacity reductions once its aluminum industry lost its international competitiveness. And it must be said that Europe does not face quite the same high energy costs as Japan. Nevertheless, the Western European policy of maintaining most of the region's existing aluminum capacity—even when the industry in particular nations is no longer internationally competitive—may be one of dubious economics that imposes costs on Western European economies and the rest of the world.

NOTES

1. The aluminum industry of Yugoslavia is omitted because its political and economic system is so different from the rest of Western Europe that its description would take the chapter far afield. Thus, in the various tables, Yugoslavian aluminum production and consumption, which are often included in the figures for Western Europe, are deducted.
2. This percentage is computed from data in table 3-1.
3. Production and consumption data from table 3-1 are used throughout this section except where otherwise noted. For more information, the reader is referred to the source shown on table 3-1.
4. The share of state-owned companies in total national capacity is as follows: Austria, 88 percent; the Federal Republic of Germany, 43 percent; and Italy, 88 percent. The French aluminum industry, which was nationalized in 1981, is totally state owned. (See also the subsequent discussion on individual countries.)
5. For discussions of the fabricating industries of the Federal Republic of Germany, see Meinhart Forster, *Struktur und Risiken der Deutschen Nichteisen-Metallversorgung* (Structure and Risks of the German Supply of Nonferrous Metals) (Hamburg, Verlag Weltarchiv GmbH, 1976) pp. 117–118. For the United Kingdom, see Kenneth Warren, *Mineral Resources* (New York, Halsted Press, 1973) p. 198. For Italy, see G. A. Baudart, "Das italienische Programm fuer die Entwicklung einer eigenen Aluminium versorgung" (Italy's Program for Domestic Supply of Aluminum), *Aluminium* vol. 48, no. 11 (1972) p. 767.

6. Table 8-6 in Carmine Nappi, *Commodity Market Controls: A Historical Review* (Lexington, Mass., and Toronto, Lexington Books and D. C. Heath and Company, 1979) p. 133, indicates that power costs amounted to 13 percent of total costs in 1970 and 21 percent in 1977.
7. Rolf Escherich, "Aluminiumerzeugung in Deutschland" (Aluminum Production in Germany), *Metall* vol. 23, no. 1 (1960) p. 56; Warren, *Mineral Resources*, p. 199; and Baudart, "Das italienische Programm," p. 767.
8. Sterling Brubaker, *Trends in the World Aluminum Industry* (Baltimore, Md., Johns Hopkins Press for Resources for the Future, 1967) pp. 129–131.
9. Restrictions were partially lifted in 1948 and totally abandoned in 1952. For a discussion of this period in the evolution of the German aluminum industry, see "Vereinigte Aluminium-Werke-Aktiengesellschaft—50 Jahre deutsche Huettenaluminiumindustrie" (VAW—50 Years of the German Aluminum Industry), *Aluminium* vol. 43, no. 6 (1967).
10. Ernst Rauch, *Geschichte der Huettenaluminiumindustrie in der westlichen Welt, 1962;* Curt Freiherr von Salmuth, *Die Aluminiumindustrie der Welt* (The Aluminium Industry of the World) (Frankfurt/Main, Giulini, 1957); Lenore Ernst, "Die Huettenaluminiumproduktion der Welt im Jahre 1969" (World Primary Aluminum Production in 1969), *Aluminium* vol. 46, no. 4; and Rolf Escherich, "Aluminiumerzeugung in Deutschland" (Aluminum Production in Germany), *Metall* vol. 23, no. 1. These articles are the sources for this and the next paragraphs.
11. "Neue Projekte der Aluminiumindustrie, Bundesrepublik Deutschland" (New Projects of the Aluminum Industry, Federal Republic of Germany), *Aluminium* vol. 44 (1968): no. 2, p. 141; no. 5, p. 338; and no. 9, p. 530; vol. 45 (1969): no. 3, p. 200; and no. 6, p. 339.
12. Lenore Ernst, "Die Huettenaluminiumproduktion der Welt im Jahre 1973" (World Primary Aluminum Production in 1973), *Aluminium* vol. 50, no. 4 (1974) p. 313.
13. Rolf Escherich, "Die Aluminiumindustrie an der Wende der Jahre 1974/75" (The Aluminum Industry at the Turn of the Year 1974/75, Federal Republic of Germany), *Aluminium* vol. 51, no. 1 (1975) p. 51.
14. "Unternehmen und Verbaende—Metallgesellschaft AG, Frankfurt/Main" (Companies and Associations—Metallgesellschaft AG, Frankfurt/Main), *Aluminium* vol. 52, no. 5 (1976) p. 353; "Unternehmen und Verbaende—Leichtmetallgesellschaft mbH" (Companies and Associations—Leichtmetallgesellschaft mbH), *Aluminium* vol. 52, no. 8 (1976) p. 535; "Unternehmen und Verbaende—Gebr. Giulini GmbH" (Companies and Associations—Gebr. Guilini GmbH), *Aluminium* vol. 52, no. 2 (1976) p. 165.

15. G. A. Baudart, *Die franzoesischen Aluminiumproduzenten—ihre Geschichte und Entwicklung* (The French Aluminum Industry—Its History and Development) (1966) p. 471; J. Boquentin, "Frankreichs Aluminiumindustrie baut ihre Stellung auf den internationalen Maerkten aus" (The French Aluminum Industry Strengthens its Position on International Markets), *Metall* vol. 19, no. 5 (1965) p. 490.
16. See Baudart, *Die franzoesischen Aluminiumproduzenten,* for a discussion of the Greek smelter; for the Netherlands and Cameroon smelters, see "Company Report of Pechiney Ugine Kuhlmann S.A. for 1971"; for the Netherlands smelter, see *Aluminium* vol. 48, no. 11 (1972) p. 780.
17. "Company Report 1971."
18. Lenore Ernst, "Die Huettenaluminiumproduktion der Welt im Jahre 1976" (World Primary Aluminum Production in 1976), *Aluminium* vol. 53, no. 14 (1977) p. 288; O. Bes de Berc, "The Aluminium Industry at the Turn of the Year 1977/78, France," *Aluminium* vol. 54, no. 1 (1978) p. 45; "Unternehmen und Verbaende—Pechiney Ugine Kuhlmann" (Companies and Associations—Pechiney Ugine Kuhlmann), *Aluminium* vol. 55, no. 10 (1979) p. 708; "Unternehmen und Verbaende—Aluminium Pechiney (Companies and Associations—Aluminum Pechiney), *Aluminium* vol. 56, no. 1 (1980) p. 135; C. Guinard, "The Aluminium Industry at the Turn of the Year 1982/83, France," *Aluminium* vol. 59, no. 1 (1983) p. 38; "Neue Projekte der Aluminiumindustrie, Frankreich" (New Projects of the Aluminum Industry, France), *Aluminium* vol. 59, no. 9 (1983) p. 720; "Unternehmen und Verbaende—Pechiney Ugine Kuhlmann" (Companies and Associations—Pechiney Ugine Kuhlmann), *Aluminium* vol. 59, no. 10 (1983) p. 806.
19. *Business Week* (February 20, 1984) pp. 54 and 57.
20. Warren, *Mineral Resources,* pp. 198–199; Dennis Swann, *Restrictive Business Practices: Studies on the United Kingdom of Great Britain and Northern Ireland, the United States of America and Japan. Part One. United Kingdom of Great Britain and Northern Ireland* (Geneva, United Nations Conference on Trade and Development [UNCTAD] E.73.II.D.8, 1973) p. 30.
21. Details are provided in BAC's company report for 1981. See "Unternehmen und Verbaende—British Aluminium Co., Ltd." (Companies and Associations—British Aluminium Co., Ltd.) *Aluminium* vol. 58, no. 8 (1982) pp. 506 and 507.
22. "Wirtschaftsnachrichten—Grossbritannien" (Business News—Great Britain), *Aluminium* vol. 58, no. 12 (1982) p. 757; "United Kingdom, Merger of Alcan (UK) and British Aluminium Completed," *International Bauxite Association (IBA)* vol. 8, no. 3 (Jan.–Mar. 1983) p. 12.
23. Ernst, "Die Huettenaluminium. im Jahre 1973," *Aluminium,* vol. 50, no. 4, p. 313; vol. 51, no. 4, p. 315; vol. 52, no. 4 (1974) p. 281.

24. Ernst, "Die Huettenaluminium. im Jahre 1969," p. 333; Lenore Ernst, "Die Huettenaluminiumproduktion der Welt im Jahre 1970" (World Primary Aluminum Production in 1970), *Aluminium* vol. 47, no. 4 (1971) p. 291.
25. The plans to construct these smelters are discussed in Baudart, "Das italienische Programm," p. 768.
26. G. Callaioli, "The Aluminium Industry at the Turn of the Year 1976/77. Italy," *Aluminium* vol. 53, no. 1 (1977) p. 41.
27. Ernst, "Die Huettenaluminium. im Jahre 1973," p. 313. MCS took a 94 percent share of Alumetal S.p.A. with smelters at Bolzano, Mori, and Fusina; Montedison took a 6 percent minority share of Alumetal. Alusuisse and MCS each took a 50 percent share of Sava.
28. Laenderkurzberichte, "Italien Auseinandersetzung um Sanierungsplan," *Aluminium* vol. 58, no. 11 (1982), p. 962.
29. "Unternehmen und Verbaende—Sava-Alluminio S.p.A." (Companies and Associations—Sava-Alluminio S.p.A.), *Aluminium* vol. 59, no. 1 (1983) p. 95. The central issue of the reorganization plan, cooperation with Alusuisse, was not settled by the end of 1983; see Laenderkurzberichte, Italien. "Staatliche Hilfe fuer den Aluminiumsektor" (State Assistance for the Aluminum Sector), *Aluminium* vol. 59, no. 6 (1983) p. 404.
30. von Salmuth, *Die Aluminiumindustrie der Welt,* pp. 95–96.
31. "Norwegen baut seine Aluminiumindustrie standig aus" (Norway Is Steadily Expanding Its Aluminum Industry) *Metall* vol. 22 (1968) p. 487; "Norwegens Aluminiumindustrie mit Rekordergebnissen" (Record Results of Norway's Aluminum Smelters), *Metall* vol. 23 (1969) p. 260.
32. See the details of the planned expansion of the Husnes smelter in "Newe Projekte der Aluminiumindustrie. Norwegen. Produktionserweiterung bei Sør Norge Aluminium" (New Projects in the Aluminum Industry, Norway. SørNorge Aluminum Plans for Expansion of Production Capacity), *Aluminium* vol. 56, no. 7 (1980) p. 506. For the Norwegian energy policy, see H. Sandvold, "Energieangebot und Standort der Aluminiumhuetten im skandinavischen Raum" (Energy Supply and Location of the Scandinavian Aluminum Smelters), *Aluminium* vol. 51, no. 12 (1975) p. 807; "Zur Situation der metallurgischen Industrie Norwegens" (The Situation of Norway's Metallurgical Industry), *Aluminium* vol. 55, no. 10 (1979) pp. 696 and 697.
33. See for 1978, "Laenderkurznachrichten. Norwegen" (Country News, Norway), *Aluminium* vol. 54, no. 11 (1978) p. 741.
34. H. Sandvold, "The Aluminum Industry at the Turn of the Year 1982/83. Norway," *Aluminium* vol. 59, no. 1 (1983) p. 51.
35. *American Metals Market* (March 7, 1984), p. 4.

36. G. A. Baudart, "Aluminium in Island—ein Projekt im hohen Norden" (Aluminum in Iceland—A Project High Up in the North), *Aluminium* vol. 42, no. 4 (1966) pp. 283–285; Lenore Ernst, "Die Huettenaluminiumproduktion der Welt im Jahre 1970" (World Primary Aluminum Production in 1970), *Aluminium* vol. 47, no. 4 (1971) p. 292.
37. "Wirtschaftsnachrichten. Island" (Business News. Iceland), *Aluminium* vol. 59, no. 4 (1983) p. 246.
38. G. A. Baudart, "Entstehung einer griechischen Aluminiumindustrie" (Development of a Greek Aluminum Industry), *Aluminium* vol. 43, no. 2 (1967) pp. 145–146.
39. Lenore Ernst, "Die Huettenaluminiumproduktion der Welt im Jahre 1972" (World Primary Aluminum Production in 1972), *Aluminium* vol. 49, no. 4 (1973) p. 23.
40. "Laenderkurznachrichten. Spanien" (Country News. Spain), *Aluminium* vol. 56, no. 8 (1980) p. 567; Gerry Butcher, *Aluminium: The International Perspective* (London, Financial Times Business Information, 1982) p. 78.
41. Butcher, *Aluminium: The International Perspective,* p. 75; United Nations, Centre on Transnational Corporations, *Transnational Corporations in the Bauxite/Aluminium Industry* (New York, United Nations, 1981), pp. 80, 81, 83, and 85. Plans for restructuring the Spanish aluminum industry provide for a rising share of the state holding company INI; see "Wirtschaftsnachrichten. Spanien" (Business News. Spain), *Aluminium* vol. 59, no. 9 (1983) p. 271; "Spain INI's plans for restructuring," *IBA* vol. 9, no. 2 (Oct.–Dec. 1983) p. 14; "Spanish aluminium industry reorganizes," *IBA* vol. 9, no. 3 (Jan.–Mar. 1984) p. 11.
42. Organisation for Economic Co-operation and Development, *Aluminum Industry: Energy Aspects of Structural Change* (Paris, OECD, 1983) p. 24.
43. Ibid., p. 112.
44. Ibid., p. 120.
45. Norbert Eickhof and Margrit Prohaska-Reichenbacher, *Die Leitungsgebundene Energiewirtschaft in Frankreich* (Bamberg, Universitaet Bamberg, Volkswirtschaftliche Diskussionsbeitraege, 1982) pp. 18 and 21.
46. Helmut Groener, "Wettbewerbliche Ausnahmebereiche im GWB: Das Beispiel der Elektrizitaetsversorgung" (Exemption of the Electric Power Industry from German Cartel Law), in H. Cox, ed., *Handbuch des Wettbewerbs* (Competition Handbook) (Munich, C. H. Beck, 1981) pp. 421–455.
47. This issue has been discussed by German aluminum companies and the Federal Ministry of Economics; see Bundesministerium fuer Wirtschaft, "Aufzeichnung. Betr. Strompreisproblem stromintensiver Produktionen" (Note on Problems of Electricity Pricing for Energy Intensive Productions) (1983) pp. 3–4.

48. Litigation between Hamburger Aluminium Werk (HAW) and the regional power company was settled by an out-of-court agreement between both parties, which led to an increase in the electricity tariff of 40 percent. See "Unternehmen and Verbaende—HAW Hamburger Aluminium Werk GmbH" (Companies and Associations—HAW Hamburger Aluminiumwerk GmbH), *Aluminium* vol. 58, no. 8 (1982) p. 507.
49. "Laenderkurznachrichten, Bundesrepublik Deutschland. Huette Ludwigshafen produziert vorlaeutig weiter" (Country News. Federal Republic of Germany. Ludwigshafen Smelter Temporarily Rescued), *Aluminium* vol. 59, no. 2 (1983).
50. Resource Strategies, Inc., "Electric Power Rates for the Aluminum Industry in the Pacific Northwest," mimeo prepared for the Bonneville Power Administration (Philadelphia, November 1984) p. 38.

4

JAPAN: A SUNSET INDUSTRY

AKIRA GOTO

The Japanese aluminum industry represents a remarkable story of the expansion and decline of a major industry. The expansion was a marked one; in two postwar decades the second largest national aluminum industry was created. In the late 1970s, there was a complete reversal, and one decade later, three-quarters of the 1973 capacity had been closed.

The precipitating event in this reversal was the sharp increase in the price of electricity in Japan to a level vastly higher than in the other five countries in this study. The high price reflected Japan's reliance on oil to generate electricity, combined with the fact that 99 percent of that oil must be imported. Japan was thus fully exposed to the two oil crises of the 1970s, which sharply raised the costs of aluminum smelting and made the Japanese aluminum industry internationally uncompetitive. Since then, aluminum imports have increased dramatically, offsetting the decline in domestic production and capacity.

The focus of this chapter is the drastic change in international competiveness of the Japanese aluminum industry and the response of the industry and the government to that change. The notion is widespread that industrial policy has played an essential role in the overall success of the Japanese economy. Yet its contribution has been exaggerated while at the same time the importance of the intense competition that has characterized Japanese industries tends to be overlooked. Industrial policy was responsible to some

Akira Goto is professor of economics at Seikei University.

degree for the success of such industries as shipbuilding and computers—at least in their early stages of development.[1] Thus, a key question to be answered in this chapter is how a public policy that was so successful in promoting the growth of some industries is now addressing an industry in decline because of the loss of its international competitiveness.

The first section of the chapter provides a brief history of the Japanese aluminum industry, an examination of its demand characteristics, and a description of its market structure. The next section, which is divided into three subsections, examines in detail the change in international competitiveness of the industry.

The first subsection examines the position of the industry prior to the first oil crisis. At that time, electricity costs were already considerably higher in Japan than in other major aluminum-producing countries. In addition, two years before the first oil crisis, the yen was revalued from 360 yen per dollar to 308 yen per dollar. Japan then moved to a flexible exchange rate system, and the yen began to float, reaching 265 yen per dollar in 1973. The revaluation of the yen considerably reduced the price of imported aluminum and had a major impact on the domestic industry. The changes in exchange rates made the industry only marginally internationally competitive. The telling blows to its competitiveness, however, were the impact of the two oil crises, which is described in the second subsection, and the resulting sharp rise in the already high price of electricity, as discussed in the third subsection.

Public policy toward the aluminum industry in Japan (described in the final section of the chapter) has been different from that of European countries where aluminum companies have received subsidies directly or indirectly in the form of reduced electricity rates from nationalized power companies. Policy measures in Japan were either responses to the immediate necessity of giving the industry breathing room, that is, time to adjust to new conditions; or long-run, if not also farfetched, efforts such as the encouragement of research on and development of a new smelting process to reduce smelter costs. Little was done to ensure a certain amount of domestic capacity through various subsidies. Yet, this does not mean that public policy toward the aluminum industry in Japan was based entirely on a free trade, free market philosophy. Japanese public policy is the product of various factors and forces that include the characteristics of the electricity market, the size and organization of the aluminum industry, and concern over potential and actual trade conflicts.

THE ORGANIZATION OF THE JAPANESE INDUSTRY

History

The aluminum smelting industry in Japan began in the 1930s, roughly thirty years after the establishment of the aluminum fabricating industry. Yet despite its late start, the aluminum industry grew rapidly. Domestic production peaked in 1943 at over 100,000 tons; as might be expected, more than 80 percent of the demand came from the military.

Three producers survived the turmoil of World War II: Nippon Light Metal, Showa Denko, and Sumitomo Chemical. In 1950, Nippon Light Metal accounted for about 65 percent of domestic production, with Showa and Sumitomo sharing the remainder almost equally.[2] Quite soon, however, high rates of growth in demand induced successive entries into the industry. In 1963, Mitsubishi Chemical started up, followed by Mitsui Aluminum in 1968 and Sumikei Aluminum in 1976. As a result, by 1980 Nippon Light Metal's market share had declined to 18.5 percent.

Japanese production exceeded 1 million tons in 1972, surpassing Canada's to make Japan the second largest aluminum-producing country among the market economies.[3] Domestic production reached its peak in 1977 with 1.2 million tons, then dropped sharply to 226,000 tons in 1985. Imports, on the other hand, rose from 466,000 tons in 1977 to 1.4 million tons in 1985 (figure 4-1). It is just such a drastic shift from domestic production to imports that underlies the characterization of the industry in the opening paragraphs of the chapter.

Most of the demand in prewar years was for military use and household utensils. In the 1950s through 1970s, demand from the transportation and building and construction sectors grew rapidly, and these two sectors soon accounted for more than 50 percent of total aluminum demand. Currently, beverage cans and electronics equipment are the two fastest-growing sources of demand. Now, however, the rate of growth of demand has slowed considerably. From 1961 to 1970, consumption increased from 183,054 tons to 886,984 tons, almost a fivefold increase. From 1971 to 1980, it increased from 1,003,905 tons to 1,640,853 tons, less than a twofold increase over the decade. This slowdown in the growth of demand was partly a result of Japan's sluggish economy. Yet even though the rate of growth has declined since the 1960s, Japan's potential for

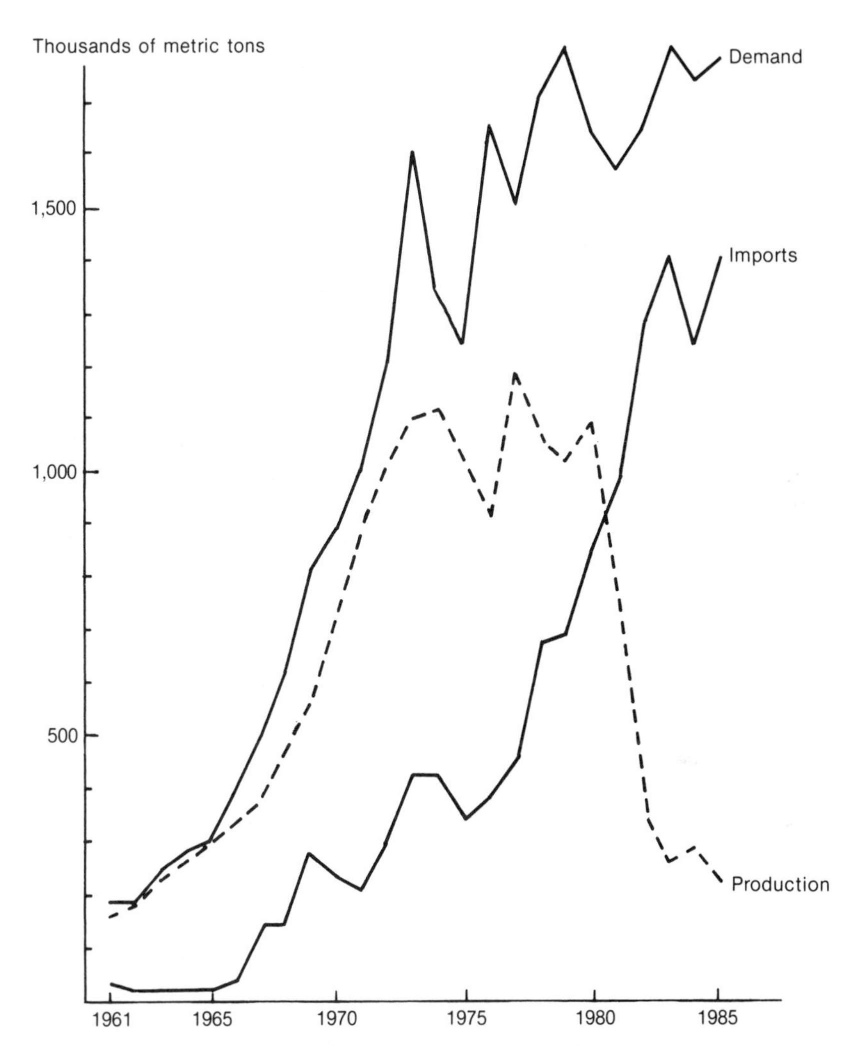

Figure 4-1. Trends in demand and production. *Source:* Ministry of International Trade and Industry, *Yearbook of Mining, Non-Ferrous Metals, and Product Statistics* (in Japanese) (Ministry of Finance, Printing Bureau, various years).

growth in demand for aluminum is still high relative to other developed countries.[4]

Market Structure

The concentration ratio in the aluminum smelting industry declined as three firms entered the market in the 1960s and 1970s. Table 4-1 shows that the dominant firm, Nippon Light Metal, steadily lost its market share. (The same pattern can be observed in the United States in such industries as steel, cans, and cigarettes during the interwar period [1918–1941].) As a result of such entry, Japan, with six producers, had more firms in its aluminum industry than any other country except the United States. In addition, two other firms, Furukawa Aluminum and Kobe Steel—the two largest aluminum fabricators—had plans in the early 1970s to integrate their operations backward and enter the smelting state, although these plans were subsequently abandoned.

Entry was encouraged by the extremely high rate of growth in demand. Aluminum smelting is capital intensive, and the minimum efficient scale is large relative to the size of the market. According to the estimates of the Ministry of International Trade and Industry (MITI), the minimum efficient scale was 100,000 tons in annual capacity in 1965, although as chapter 1 points out, there are probably some scale economies above that capacity. The size of the Japanese market at that time was about 300,000 tons, making the ratio of the minimum efficient scale to the size of the market more than 30 percent.[5]

In addition, by 1965 most of the good smelter locations—that is, those with hydropower generating low-cost electricity—were already in use. Therefore, the power alternatives for new aluminum producers were limited to building a thermal power plant jointly with a power company, purchasing electricity from a power company, or using a mix of the two.

Both the size required for an efficient plant and the high-priced power situation created an industry with very high barriers to entry. Yet entry continued despite the barriers because there was an expectation of continued high rates of growth of demand. This expectation was more than confirmed in the 1960s by growth in the domestic consumption of primary aluminum at 20.8 percent annually. In 1963 the demand for aluminum grew an amazing 39.6 percent from the previous year. By 1971 Japanese aluminum consumption exceeded 1 million tons. The minimum efficient scale, which as noted

Table 4-1. Aluminum Market Shares,[a] by Company, for Selected Years, 1950–1980 (percentage)

Company	1950	1955	1960	1965	1970	1975	1980	1985
Nippon Light Metal	64.9	48.5	47.9	38.8	28.6	29.1	18.5	25.1
Sumitomo Chemical[b]	17.8	20.2	21.2	26.3	25.8	24.0	25.4	18.8
Sowa Denko[b]	17.3	31.3	30.9	25.3	24.7	19.2	14.1	10.0
Mitsubishi Chemical[b]				9.6	19.9	20.1	20.8	18.2
Mitsui Aluminum					1.0	7.6	12.2	27.9
Sumikei Aluminum							9.0	

Source: For 1950 and 1955: Fair Trade Commission, *Industrial Concentration in Japan* (in Japanese) (Tokyo, Toyo Keizai Shinposha, 1957); 1960–1980: Japan Aluminum Federation, *Light Metal Statistics in Japan* (in Japanese) (Tokyo, various years).

[a] Note that the figures are each company's share of total domestic production.

[b] The aluminum divisions of these chemical companies were eventually separated from the parent chemical companies and became independent, although still affiliated, companies. Sumitomo Chemical's aluminum division became Sumitomo Aluminum Smelting, Showa Denko's aluminum division became Showa Aluminum Smelting, and Mitsubishi Chemical's aluminum division became Mitsubishi Light Metal, later Ryoka Light Metal.

earlier was estimated at 100,000 tons in 1965, changed little because the technology of the electrolytic process remained largely unchanged.[6] Thus, ten plants of the minimum efficient scale could have been supported by the 1971 domestic consumption. Furthermore, the number that could have been supported became even larger as the market continued to expand in the 1970s, to more than 1.5 million tons.

Another reason entry continued in the industry was the conservative investment behavior of the established producers of primary aluminum. As discussed later in this chapter, the production costs of Japanese smelters were high relative to their counterparts in North America, mainly because of the high price of electricity. Also, imports had been garnering an increasing share of the Japanese market ever since the importing of aluminum was liberalized in 1961. Thus, Japanese smelters were under competitive pressure to keep their production costs as low as possible in order to compete with imports. (There was no possibility from the beginning for the Japanese aluminum industry to become an exporter.) As is typical in a capital-intensive industry, it was vitally important for aluminum producers to maintain a high operating ratio to keep production costs low. According to MITI estimates, production costs, vis-à-vis full capacity, became 5 percent higher with a 90 percent operating ratio, 11 percent higher with an 80 percent operating ratio, and 19 percent higher with a 70 percent operating ratio.[7] Therefore, the investment policy of the existing primary producers was to expand in keeping with the pace of growth in domestic demand. But partly because of the concern to keep operating ratios high and partly because actual demand grew much faster than their expectations, the estimates of the growth of demand by existing producers were always below the actual growth rate of aluminum consumption. Thus, opportunities were created to enter the industry, and these opportunities were exploited by three entrants. (And had the oil crises not taken place, there would have been two more entrants.)

The absence of forward vertical integration into fabricating, which is typical of the aluminum producers of other industrialized countries, has important implications for public policy in Japan. Japanese fabricators began operating around the turn of the century using imported aluminum from the United States and Europe. Yet because of the competition from low-priced imports, a lack of sophisticated technology, the less developed state of electricity generation in Japan, and the modest size of the market, all of the several attempts to begin smelting in the 1910s and the 1920s failed. By the 1920s, a well-established fabricating industry existed as divisions or subsidiaries of mining and metal-producing firms. Attempts to establish aluminum smelting operations continued, however (most of the attempts were made by chemical firms). They finally succeeded in the mid-1930s.

After the war the Japanese aluminum industry developed without a major attempt to integrate vertically from either the fabricating or the smelting side. With demand growing rapidly, primary producers had little incentive to integrate forward because they could exploit the intense competition among fabricators. (In 1963 there were thirty-four fabricators. From 1955 through 1964 fourteen firms entered the market, and from 1965 through 1974 another thirty firms entered the fabricating sector.) Rather than extending their operation into the fabricating stage, primary producers with market power through the sales of aluminum ingot and without the resources to market fabrications were better off concentrating on smelting. In addition, all of the producers had fabricators in which they had capital participation. Through this quasi-vertical integration, producers could secure a stable outlet for their aluminum without incurring the risk and costs associated with integration into the fabricating and marketing stages.

Large fabricators, on the other hand, had an economic rationale to integrate backward into the smelting stage because their consumption of primary aluminum increased rapidly in the 1960s, and occasionally there were shortages. In the 1970s three fabricators made plans to enter the smelting side of the industry. Before they could carry out their plans, however, the oil crises occurred, and the price of electricity rose sharply. Except for one fabricator—Sumitomo Light Metal—all such entry plans were abandoned. (Sumitomo Light Metal started smelting in 1976 by establishing a subsidiary, Sumikei Aluminum. That company survived only seven years.)

After the oil crises, the industry's situation changed dramatically. Fabricators had no incentive to integrate into the smelting stage because the price of electricity rose very sharply—from an already high level. On the other hand, primary producers—particularly Nippon Light Metal—pursued a strategy of forward integration as a means of survival. Nippon merged with its fabricator subsidiary in 1974; today, the company is gradually shifting its main line of business from smelting to fabricating and the end-product stage.

Aluminum Producers and Business Groups

Producers of primary aluminum did not diversify into products other than aluminum. To be sure, these firms were themselves the consequence of the diversification of large chemical firms. Yet the reluctance of the resulting primary producers to engage in product diversification can be better explained by technological, marketing, and organizational reasons.[8] Primary producers were oriented in their technology toward processes rather than products. They lacked an elaborate marketing system for product diversification. In addition, their organization was departmentalized along functional lines, a structure that tended to discourage any searching for opportunities to diversify into new product groups. Japanese primary producers were in essence carrying all of their eggs in one basket, and the basket was made of aluminum.

Another important feature of the aluminum industry in Japan was that the Japanese primary producers were all members of one of the six major business groups.[9] Japanese business groups are loose coalitions of firms pursuing their mutual interests through a system that coordinates their decisions. Unlike pre–World War II *Zaibatsu* or groups in European countries, member firms of a Japanese business group are not under the control of a single holding company. All the members are formally equals, although trading companies and banks are usually more dominant in group decisions. The absence of a central decision-making unit makes the relationship among member firms in a group rather loose. Although the six major groups are significant in the Japanese economy, there are many large, important firms that do not belong to any of the six groups.

Group membership for the primary aluminum producers can be explained by several factors. First, because aluminum smelting is extremely capital intensive, the capital requirements of a smelter built to an efficient scale are substantial. According to MITI estimates, a 100,000-ton capacity smelting plant (the minimum efficient-scale capacity) would have cost about 26 billion yen ($72.2 million) in 1968.[10] In addition, entrants would have required about 10 billion yen ($27.7 million) for an alumina plant with a 200,000-ton capacity and another 10 billion yen ($27.7 million) for a share in a power plant built jointly with the power company.[11] To raise this amount of capital (46 billion yen or $127.6 million) is not easy, especially in the imperfect capital markets that have characterized the Japanese economy. Business groups have an obvious advantage in raising such capital; where there is a promising yet risky investment opportunity, groups can share the risks. This risk sharing is particularly important when a large investment is required. Groups are also useful when a product requires a combination of different types of technology. The coordination and organization of members is done through long-established channels of information with mutual trust secured by financial linkages and interlocking directorates. Hence, the transaction costs associated with joint investment are less when group members invest jointly then if an ad hoc joint venture were formed.

Second, as mentioned earlier, primary producers are not integrated forward in Japan. As a result, primary aluminum is sold in market transactions to fabricators. To keep their operating ratios high, primary producers need stable outlets for their aluminum. In such a situation, group affiliation becomes an alternative to vertical integration. When a fabricator or its parent company is a group member, the newly formed primary producer of this group is assured a stable outlet without investing either in the fabricating stage or in the establishment of a sales organization. Thus group affiliation offers smelting firms a stable outlet for their product at a low cost.

Third, it is sometimes argued that group affiliation works as insurance. When one member suffers heavy losses for some time and is on the verge of bankruptcy, other members of the same group try to help the firm by providing loans, sharing losses, and accepting workers into their firms. There seems to be a tacit understanding among group members that they will help one another when necessary. For example, when coal mining firms that were group members encountered critical conditions in the 1950s as a result of the switch from coal to oil, other group members helped them to reorganize through the measures mentioned above. Such practices appear to reduce the pain and cost associated with adjustments in the industrial structure of an economy. Primary aluminum producers received the same kind of assistance from other group members in adjusting to the oil shocks of the 1970s.

CHANGING INTERNATIONAL COMPETITIVENESS

Before the Oil Crises

From its inception, the Japanese aluminum industry has confronted intense competition from abroad. During the first three decades of this century, several attempts to start the production of primary aluminum failed, although domestic consumption was steadily increasing, because of the high production costs relative to the price of imported aluminum. Costs were high largely because there was no domestic bauxite, electricity was expensive, and there was a lack of knowledge of the technology. Although gradually bauxite became available from abroad and the necessary technology, including the electrolytic process, was acquired and developed, the price of electricity remained high to threaten the cost competitiveness of production in Japan. Despite this constraint, primary aluminum production began in Japan and expanded to create a major industry, but its international competitiveness was always fragile.

The oil crises of the 1970s finally destroyed the delicate balance that had allowed primary production to be carried out in Japan. Yet it is important to understand that from the beginning the Japanese aluminum industry was in a weak position vis-à-vis imports. Even prior to the oil crises, a series of events substantially eroded the competitiveness of the Japanese aluminum industry.

Table 4-2 is a comparison of electricity prices for aluminum smelting in major aluminum-producing countries in 1957. The price of electricity in Japan was about four times as high as that in Canada, and more than twice as high as the average in the United States. With electricity prices at this level, the total production cost of aluminum in Japan was estimated to be 197,900 yen ($550) per ton in 1962.[12] On the other hand, the import price for aluminum at this time was $485–$496 per ton. Adding the handling fee, marketing expenses, and a 10 percent tariff would increase the cost of imported aluminum to 195,100–199,400 yen ($540–$555) per ton. This suggests that domestic production costs could be comparable to the price of imported aluminum with a tariff rate of 10 percent. (At a 15 percent tariff, the price of imported aluminum would be 203,800–208,400 yen [$566–$580] per ton, higher than the domestic

Table 4-2. Electricity Prices to Aluminum Smelters in Various Countries, 1957 (mills per kilowatt-hour)

Country	Price
Japan	7.78
United States	1.94–4.44
Canada	1.94
West Germany	3.75–4.17

Source: Sangyokozo Chosakai, *Industrial Structure in Japan,* vol. 3 (in Japanese) (Tokyo, Tsusho Sangyo Kenkyusha, 1964).

production cost.) A 10 percent rate thus was a necessary condition for the survival of the domestic smelters.

The tariff rate was set at 15 percent when the importation of aluminum was liberalized in June 1961, and it remained at that level until March 1964. It was reduced to 13 percent in April 1964 and further reduced to 11.4 percent in July 1968. Thus, although it was reduced gradually, the tariff rate for aluminum ingot remained above 10 percent. This apparently allowed Japanese producers to compete successfully with imports during the 1960s.

In addition to tariff protection, Japanese smelters had the advantage of being near a large and rapidly expanding market. By maintaining close contacts with fabricators and other users through group affiliation, they could meet buyer needs quickly and efficiently. Fabricators as well were eager to establish a close relationship with producers that could be stable and dependable suppliers. The advantage of the Japanese producers was further reinforced by the rather tight world aluminum market at that time. North American smelters were more concerned about fulfilling their commitments to their old customers than expanding their sales. With the help of these factors, Japanese smelters could compete effectively against the imports, although their edge was admittedly slim and rested on a 10 percent tariff.

Two events that took place in 1971 changed this delicate balance and undermined the competitiveness of the Japanese producers. The first was the reduction of tariff rates as a result of the Kennedy round, including the application of preferential tariff rates to a part of the imports from the developing countries. The tariff rate was reduced to 10.6 percent in January 1970, then reduced further to 9.8 percent in January 1971, and finally, reduced again to 9 percent in 1971. In addition, preferential tariff rates (4.5 percent) were applied to imports from the developing countries. Imports under this preferential tariff amounted to 15 percent of the total aluminum imports in 1971 and 21 percent in 1972. Although it is not easy to assess the impact of the tariff reduction, the previously cited MITI cost estimates and calculations suggest that tariffs functioned effectively as a barrier to protect the domestic producers. This barrier was lowered substantially in the early 1970s.

The revaluation of the Japanese yen was an even more significant event than the tariff reductions. The exchange rate of the Japanese yen to the U.S. dollar had been pegged at 360 yen per dollar after World War II and had remained at that level. Japan moved to the flexible exchange rate system in 1973, and the dollar declined to a value of around 270 yen.

This sharp revaluation of the yen had two effects on the Japanese aluminum smelting industry. It reduced the price of imported aluminum within the Japanese market. On the other hand, it also reduced the production costs of domestic smelters because the price they paid for imported materials decreased. Yet the cost savings of domestic smelters were much smaller than the decrease in the market price of the imported aluminum.

A simple calculation illustrates the point. The ratio of the cost of materials to the total production cost for domestic aluminum smelters was estimated to be around 35 percent. As mentioned above, from 1971 to 1973 the yen was revalued from 360 yen per dollar to about 270 yen per dollar—approximately a 25 percent increase. Other things being equal, this would reduce the total production cost of domestic smelters about 8 percent at most.[13] The revaluation, however, would also lower the market price of imported aluminum by a full 25 percent because contracts to import aluminum were usually quoted in terms of dollars.

According to Tashita,[14] the price of imported aluminum (C.I.F. basis) fell from 180,000 yen ($571) per ton in July 1971 to 129,000 yen ($460) in April 1973, a decrease of about 28 percent in terms of yen; domestic production costs, on the other hand, decreased on 2 percent. As a result, the market price of imported aluminum, which was the sum of the C.I.F. price, handling and marketing expenses (5–8 percent of the C.I.F. price), and tariff charges (9 percent in general but 4.5 percent for those suppliers to which the preferential tariff was applicable after April 1971), fell well below the production cost of the domestic smelters. The 129,000 yen ($460) import price was considerably lower than the 170,000–190,000 yen ($600–$680) per ton production cost estimated by Tashita.[15]

Consequently, imports increased sharply, reflecting this shift of competitiveness coupled with an excess supply of aluminum on the world market. In 1971 Japan imported 209,000 tons of primary aluminum; its imports increased to 293,000 tons in 1972 and 428,000 tons in 1973. Thus, revaluation of the yen, intensified by the reduction of tariffs, had a severe impact on the competitiveness of the Japanese aluminum industry even before the first oil crisis.

The Impact of the Oil Crises

The first oil crisis occurred at a time when Japanese aluminum smelters were losing their competitiveness because of the revaluation of the yen and the reduction of tariffs. The oil embargo by the

Arab countries increased the price of oil from $4.75 per barrel in 1973 to $12.05 per barrel in 1974. When the second oil crisis took place in 1979 because of the revolution in Iran, the price rose to $34.61 per barrel in 1980, more than seven times the pre–oil-crisis level. These two events were a serious blow to the Japanese aluminum industry, which depended heavily on oil to generate electricity.

When the first oil crisis occurred in 1973, the Japanese economy was in the middle of a boom, and the demand for aluminum had reached a new peak of 1.6 million tons. In addition, the world aluminum market had resumed its growth in demand with a simultaneous expansion in the other industrialized countries. Growth in the world supply of aluminum, however, was held back because of a low operating ratio in the smelters in the U.S. Northwest caused by a drought that cut down the availability of hydropower. As a result, the price of primary aluminum rose, offsetting the rise in the cost. In the latter half of 1974, however, the boom ended with the sharp increase in the price of oil that signaled the first oil crisis. The Japanese government responded by adopting a policy to restrain aggregate demand in order to counter inflation, and the Japanese economy fell into a recession. The demand for aluminum in 1974 decreased from the previous year for the first time in the post–World War II period. Japanese producers were forced to cut back their production, which caused an increase in the production costs per ton because of the low operating ratio. Production costs rose even more because of the sharp rise in the price of electricity that resulted from the oil price increase.

The entry into the market of Sumikei Aluminum caused the operating ratio of existing producers to drop even further. (This entry was based on the construction of a smelter before the recession and oil crisis.) Imports also continued to increase substantially from 1977 through 1978 because of widening cost differentials between domestic smelters and the smelters abroad. This trend of increasing imports was further accelerated by a sharp rise in the exchange rate from 296 yen in 1976 to 210 yen per U.S. dollar in 1978. Imports reached 674,000 tons in 1978, accounting for 40 percent of the domestic demand and further reducing the operating ratio of domestic smelters. From 1974 to the end of 1978, the industry's total accumulated loss was 87.9 billion yen ($451.7 million).

Yet in 1979 and 1980 the aluminum industry seemed to escape from this vicious cycle. By then, the Japanese economy had recovered from the recession. Demand for aluminum reached 1.8 million tons, the highest level ever achieved. The world economy was also recovering from the recession, which pulled up the imported price of aluminum. In addition, Japanese producers formed a depression cartel (a legal exception under the Anti-Monopoly Law) and were able to cut back production by 40 percent from September 1978 through March 1979. Under these conditions the market price of aluminum jumped from 290,000 yen ($1,490) per ton in 1978 to 508,000 yen ($2,500) in 1980. As a result, the industry earned a profit in 1979 and 1980 for the first time after the first oil crisis, reducing the accumulated loss of the industry since 1974 to 31.9 billion yen ($140.7 million).[16]

The second oil crisis, triggered by the revolution in Iran, ended this short boom abruptly. The price of oil rose from $23.07 per barrel in 1979 to $38.24 per barrel in June 1981. With this price rise, the cost of electricity for the aluminum smelters using oil as the source of electricity generation reached the almost prohibitively high level of around 16 yen (72 mills) per kilowatt-hour. Total costs exceeded 500,000 yen ($2,100) per ton for smelters using oil as the primary source of energy to generate electricity, and the ratio of electricity to total costs exceeded 50 percent.[17] In 1982 the price of imported aluminum was about 300,000 yen ($1,280), and the volume of imports exceeded domestic production for the first time. Imports as a share of domestic consumption totaled 62 percent in 1981 and 79 percent in 1982; domestic production dropped from 770,000 tons to 350,000 tons. The loss in the industry in 1981 was 69.8 billion yen ($317.4 million); in 1982 losses increased to 101.8 billion yen ($433.2 million).

It was then apparent that there was no future for the smelters using oil as the source of electricity generation. From late 1981 through 1982, five smelters were shut down. Counting the two smelters that had already been closed in 1979 and 1980, seven of fourteen smelters in Japan were closed by the end of 1982. Among the surviving seven smelters, two are currently being considered for closure.[18] One company, Sumikei Aluminum, was forced to withdraw from the industry in 1982, six years after its entry. The number of workers in the aluminum smelting industry was reduced from 15,000 in 1980 to fewer than 5,000 in 1984.

A few of the smelters have advantages that may make it possible for them to survive for a few years. For example, the Kanbara smelter of Nippon Light Metal has its own hydropower plant, whereas the Miike plant of Mitsui Aluminum has access to less expensive electricity from a nearby power plant that it established jointly with the local power company. This plant has lower costs because it uses the powder coal produced as a by-product at the nearby affiliated coal mine. The remaining smelters are trying to survive by switching from oil to coal as the source of electricity generation. Taking into consideration all of these special circumstances, domestic production will probably be about 300,000 to 350,000 tons, compared with a 1977 peak output of 1.2 million tons.[19]

There are several reasons why the Japanese aluminum industry suffered so severely. Even before the oil crises, Japanese smelters were barely competitive with imports in the domestic market because of the high prices they paid for electricity. They did have the advantage of being located near the rapidly growing market and thus were able to keep in close touch with their customers. With tariff protection, they could compete and even grow at a high rate. Also, because the world aluminum market was generally tight and controlled by the so-called big six producers, the import price of aluminum remained stable at a relatively high level. For Japanese smelters to prosper at their pre-1970 scale, all of these conditions were necessary.

But, from the late 1960s through the early 1970s, these favorable conditions disappeared one by one. First, the reduction of tariffs and the revaluation of the yen undermined the competitiveness of the Japanese smelters. Then the world and domestic markets became depressed. A number of the developing countries began aluminum smelting, and there were also many new firms entering the industry in the other industrialized countries. The rate of growth of demand first slowed and then with the recessions became negative, resulting in declines in the world price of primary aluminum. As a consequence, import prices became significantly lower than the production costs of the domestic smelters.

When the conditions surrounding the smelters became increasingly difficult, the financial position of the Japanese aluminum producers deteriorated rapidly because of their lack of diversification and forward integration. When the cost differentials between domestic and imported aluminum became substantial in the 1970s, fabricators quickly switched to cheaper imports. Affiliated fabricators also switched because they, too, had to survive. Apparently there was a limit to the protection provided by the quasi-vertical integration or group membership.

The ultimate reason for the decline of the Japanese aluminum smelting industry, however, was the high price of electricity, a problem that had afflicted the industry from its start in the 1930s.

Although other favorable conditions mitigated the severity of this problem in earlier years, it surfaced again in the 1970s when oil prices rose drastically.

The Supply of Electricity

Given its importance, the supply of electricity to the Japanese aluminum industry warrants further examination. Primary producers in Japan could obtain electricity in three ways: owning their own power plant, building the power plant jointly with the power company, or purchasing electricity from a power company. Smelters either used a single method or a mixture of them. Table 4-3 shows types and sources of electricity used by producers that fall into these three categories, as well as the sources of primary energy for generation at privately owned power plants and jointly owned power plants. In 1965 purchased electricity was the largest source; after 1965, however, five of the seven smelters that were constructed had power plants built jointly with the local power company. All five used oil as the primary generating source. (One of the remaining two plants purchased its electricity; the other has its own power plant using coal.)

Thus, the share of electricity from oil-burning power plants increased rapidly after 1965. Taking into account the fact that the dependence of power companies on oil was also high, the total amount of electricity used by smelters that was generated from oil reached 78 percent. With such a heavy dependence on oil as the source of electricity generation, the impact of a sevenfold increase in the price of oil needs little imagination. Twice in the 1970s the electricity rates of power companies rose sharply because of increases in the price of oil: in 1974, the rates for large-scale industrial users rose 84.77 percent; in 1980, the rates rose 57.71 percent. In addition, after the second oil crisis, electricity costs were said to be higher for the smelters with jointly owned power plants than for the smelters that purchased electricity from power companies because the power companies' sources of primary energy were more diversified.[20] Smelters using jointly owned power plants that burned oil were either closed or switched from oil to coal. As a result, as can be seen from table 4-3, dependence on oil had fallen by 1983, whereas dependence on coal had risen considerably.

There are several characteristics of the Japanese electric power industry, which is served by nine local monopolies and regulated by

Table 4-3. Electricity Used to Smelt Aluminum, by Type and Source (million kilowatt-hours)

Type and Source	1965	1973	1981	Feb. 1983
A. Electricity from privately owned plant	1,530 (27.5)	4,935 (27.2)	2,992 (29.5)	(83.9)
B. Electricity from jointly owned plant	1,883 (33.8)	9,203 (50.7)	5,774 (56.7)	
C. Purchased electricity	2,150 (38.7)	3,994 (22.1)	1,403 (13.8)	(16.1)
Primary energy for A + B				
Hydro	1,216 (21.8)	1,260 (6.9)	1,216 (12.0)	(19.1)
Coal and others	544 (9.7)	1,961 (10.8)	1,761 (17.4)	(42.2)
Oil	1,655 (29.8)	10,917 (60.2)	5,758 (56.8)	(22.4)

Note: Numbers in parentheses are percentages.
Source: The data in this table were provided to the author by the Japan Aluminum Federation.

MITI, that contributed to the sharp decline of the aluminum industry. First, the Electricity Industry Act sets forth three criteria for electricity: (1) rates must be determined on the basis of necessary costs, (2) a fair return on investment must be included in the total cost, and (3) there must be no rate discrimination among users. Whenever costs rise, as in the case of the oil crises, such costs are passed on to the users, and following the third principle, little relief is afforded to the extremely large-scale users like aluminum smelters. On the contrary: an increasing rate structure was introduced in 1974 to provide an incentive for energy conservation. Under this system, as a customer uses more electricity and exceeds the "standard amount of electricity," which is determined from past records of power use, an additional rate is charged for power that exceeds the "standard." Although a special contract rate system including time and seasonal discounts and further discounts for interruptible power was introduced in 1979, the steep increase of electricity rates and the change of the rate structure in a way unfavorable to large-scale users had an adverse effect on smelters that depended on purchased electricity.

Second, there are no long-term contracts between power companies and their customers. In addition, it takes only about two months for an application for a rate increase to become effective. Therefore, unlike smelters in other countries whose long-term power contracts shielded them from the impact of the abrupt oil price rise for several years, rate increases became effective very quickly in Japan.

To summarize, electricity costs for the smelters with jointly owned power plants using oil rose sharply because of the oil price rise. Electricity costs for those smelters purchasing electricity from power companies also rose sharply because the rise in the price of oil was passed on to users quickly—especially to large-scale users like aluminum smelters.

THE RESPONSE OF PUBLIC POLICY

The final section of this chapter examines the response of public policy in Japan toward a rapidly declining aluminum smelting industry. Such policy was formulated, however, in the context of the responses and adjustments made by the primary producers themselves to the industry's rapidly changing conditions.

Adjustment by the Producers

The various strategies pursued by Japanese aluminum producers fall into four categories. First, producers tried to reduce their power costs by shutting down the plants that used electricity generated from oil and concentrating their operations in smelters that used lower cost hydropower or coal to generate electricity. As a result the dependence on oil dropped considerably. Producers also reduced capacity from 1.64 million tons in 1977 to 712,000 tons in 1984.

Second, beginning in the early 1970s, all of the primary producers invested in overseas smelting projects. Consequently, there are a total of nine offshore projects completed, under construction, or in the planning stage. Imports from these projects reached 473,000 tons in 1983, accounting for about 34 percent of all aluminum imports and 26 percent of domestic comsumption. These imports are expected to increase to 740,000 tons in 1988, which would be about 35 percent of domestic consumption; domestic production by 1988 should be about 300,000 tons, or about 15 percent of domestic consumption. (The remaining 50 percent of domestic consumption is expected to be met by imports on a long-term contract or spot basis.)[21] Most of the aluminum produced domestically will be high-grade aluminum primarily for high-tech industries, including electronics. Thus, the main line of operation of Japanese aluminum producers will shift abroad.

Third, in an effort to reduce energy costs, producers also started research on a new smelting process. This effort was initiated by Mitsui Aluminum and the Institute of Chemical Technology of the Agency of Industrial Science and Technology and developed into a joint research project under the aegis of the Research Association on the New Smelting Process of Aluminum. All of the other producers and a number of engineering firms are members.[22]

The essence of this process, which is similar to that for producing steel, is to reduce the ore containing aluminum directly in the blast furnace with coke.[23] The next step is to refine the obtained aluminum-siricon or aluminum-siricon-ferrum alloy into pure aluminum. In theory, this process requires only 1,000 kilowatt-hours of electricity per ton of aluminum, and it is possible to generate more electricity than needed in this process by using carbon monoxide, which is produced as a by-product. It is also possible to use ores other than bauxite as a raw material. As chapter 1 indicates, however, there are serious technical difficulties remaining, in particular, the development of the lining of the blast furnace (the heat inside the furnace is expected to rise to 2,000°C) and the vaporization of aluminum inside the furnace. Because of these problems, it is widely believed that it will take at least six to seven years to make the process operational—if it is possible at all.

Finally, aluminum producers have begun to integrate and diversify, moving into fabrications and the production of other products. Nippon Light Metal is the most active producer following the integration strategy, absorbing affiliated fabricators and even trying to produce consumer products such as ice cream makers. Showa Aluminum is also using its production of alumina to diversify into the fine ceramics industry using alumina as its main material.

Government Actions

With the deterioration of the aluminum smelting industry's competitiveness, MITI designated it a "specific depressed industry" under the Temporary Measures Law for the Stabilization of Specific Depressed Industries, which was enacted in 1978 and remained in effect for five years. This law was followed by the Temporary Measures Law for Structural Adjustment of Specific Industries enacted in 1983 and also effective for five years.[24] Under the first law, fourteen industries were designated as depressed industries; under the second law, twenty-one industries were covered as of the end of September 1983. The aluminum smelting industry was included in both cases. All of the industries so designated are required to develop a stabilization or structural improvement plan. In carrying out the plans, which are mostly concerned with the disposition of excess capacity, various forms of government assistance are provided.

The Basic Plan for Stabilization for the aluminum industry under the 1978 law and the Basic Plan for Structural Improvement under the 1983 law included two types of governmental actions: (1) those aimed at providing the industry with emergency relief so that producers would have time to adjust to rapidly changing conditions and (2) those aimed at assisting the long-run survival of at least some domestic production.

The first group of policy measures included a number of features.

A tariff quota system and a tariff exemption system were created to finance the disposition of excess capacity. The initial stabilization plan for the industry expected to keep 1.1 million tons of domestic capacity in operation; about 500,000 tons of capacity was considered redundant. But, it was not possible to scrap a large portion of production without serious effects on the financial position of producers, mainly because, like other Japanese firms, they were heavily dependent on debt financing. Canceling large debts would have repercussions on banks, insurance companies, and other firms including parent companies that held debt as well as equity. Therefore, it was considered necessary to provide some form of assistance to aluminum firms to carry out the stabilization plan without financial casualties.

The tariff quota system, which was in effect in the industry during 1978–1979, was used to provide transitional financial assistance. Importers (including fabricators) could import aluminum—on a first-come, first-served basis—at reduced tariff rates until the amount of imports reached the amount of the demand in excess of the capacity that was planned to be maintained under the stabilization plan. After this limit or quota was reached, importers had to pay the normal tariff rates. The importers also had to pay a sum equal to their tariff savings (after deducting the handling fee) to the Association for the Promotion of the Structural Improvement of the Aluminum Industry. The association in turn paid this structural improvement fund to the primary producers. The amount a producer received was the book value of the capacity it disposed of or put into standby status, multiplied by an interest rate of 6.6 percent. This subsidy was to cover the annual interest cost of the capacity that was no longer in operation. It should be emphasized that the tariff quota system was not invented for the aluminum industry; in 1978 a total of fourteen industries were using such a system.

In 1982, after the debilitating impact of the second oil price rise, the original Basic Plan for Stabilization, based on the assumption of maintaining 1.1 million tons of capacity, was judged to be unrealistic. The new plan, the Basic Plan for Structural Improvement, was devised under the 1983 law; it involved reducing the target capacity to be maintained from 1.1 million tons to 700,000 tons, thus requiring the closing of another 400,000 tons of capacity. To achieve this goal the tariff exemption system was introduced. This system allowed primary producers to import aluminum without paying a tariff (which at the time was 9 percent of the value) up to the amount of the capacity they had closed under the structural improvement plan. This provision was in effect until 1984.

During the early 1980s, imports continued to increase while domestic production dropped even further. The volume of imports increased from 980,000 tons in 1981 to 1.29 million tons in 1982 and

1.42 million tons in 1983; domestic production, on the other hand, decreased from 770,000 tons in 1981 to 350,000 tons in 1982 and 255,000 tons in 1983. Thus, the structural improvement plan developed under the second law, which was based on the assumption of keeping a 700,000-ton domestic capacity, also came to be seen as unrealistic. At the end of fiscal year 1984, when the tariff exemption system was to be terminated, the plan was revised to dispose of (or place on a standby basis) another 360,000 tons of capacity by 1988. The tariff exemption system was extended, with a slight change, to finance the dispostion of this capacity.[25]

These schemes enhanced the willingness of aluminum producers to dispose of or put into standby their excess capacity. As a result, in 1978 and 1979, while the tariff quota system was in effect, 506,000 tons of capacity were eliminated. By 1984 another 424,000 tons of capacity had been eliminated or placed on standby under the tariff exemption system.

The most difficult problem for all of the industries covered by these laws has been the allocation of capacity reductions among firms. Conflicts over this allocation often lead to a collapse of the arrangement itself because each firm often expects other firms to cut back their capacity or leave the market entirely so that there will be more demand for a remaining firm and make it possible for that firm to survive—the classic "free rider" problem. Although there are no public records of how producers have resolved these conflicts, the options are known. For example, producers have the option of discussing and negotiating terms with each other. In addition, MITI can exercise "administrative guidance" (that is, jawboning) to persuade firms who are reluctant to dispose of their own capacity in hopes that another firm would dispose of its capacity first.

In the case of the aluminum industry, however, agreement on the allocation of the amount of capacity reduction to each firm was less difficult. The losses of the producers were already enormous, and they all shared a pessimistic view of the future of aluminum smelting in Japan. Also, the cost differentials among the smelters were significant and apparent. A consensus readily emerged to reduce the industry's capacity and to close the higher cost smelters.

It should be noted that, until the end of 1983, the subsidy that producers received under the tariff quota and tariff exemption systems was about 28 billion yen ($120.5 million using the exchange rate of the end of 1983). This subsidy is far smaller than the loss shouldered by the producers and their parent companies, which is estimated to be around 186 billion yen ($801 million using the exchange rate of the end of 1983) from 1976 to 1983. Thus, the adjustment cost to the aluminum industry was largely shared by the producers and their parent companies; the role of the government subsidy was relatively modest. The important aspect of these subsidies was not their amount, however, but their purpose. They were granted to encourage the disposition of capacity and not to maintain the current scale of operations.

There were two other measures enacted under the 1978 law that also deserve mention. The stabilization plan provided for the purchase of aluminum by the government for stockpiling. The Light Metal Stockpiling Association thus bought about 167,000 tons of primary aluminum (worth about 61.6 billion yen or approximately $265 million using the exchange rate of the end of 1983) from smelters during 1976–1983. In addition, interest rates were reduced for loans to the primary producers from the Japan Development Bank, the semigovernmental financial institution.

The second group of policy measures emphasized long-run adjustment. For example, jointly owned power plants were encouraged to switch from oil to coal. Also, subsidies and low-interest loans from the Japan Development Bank were provided to jointly owned power plants to reduce the price of electricity for the aluminum industry.

Research was also subsidized on the new smelting process described earlier. The basic research for its development had been carried out at a government laboratory (the Institute of Chemical Technology of the Agency of Industrial Science and Technology). In addition, a subsidy was granted to the Research Association on the New Smelting Process of Aluminum, the joint research effort by the smelters and several engineering firms.

The two temporary measures laws were accompanied by laws that dealt with the adjustment of employment. Under these laws, the relocation of workers from "the specific depressed industries" to other expanding industries is encouraged. For instance, subsidies are provided to a firm that employs workers from "the specific depressed industry" and to firms belonging to "the specific depressed industry" that train workers planning to leave the industry. In addition, the term of payment of unemployment insurance is extended.

Here, again, Japan's public policy toward labor, like its policy toward production capacity, is a mix of two: namely, the provision of "breathing room" (for example, extension of the unemployment insurance) and the encouragement of adjustment (for example, subsidies to encourage the relocation of workers). The country's policy emphasis, however, is on the latter. The stance of public policy toward "the specific industries" is to relocate the factor of production smoothly out of these industries and toward those that are expanding with as small an adjustment cost as possible.

Economic Evaluation

The implications of Japan's adjustment assistance policy are examined by Mayer, Lapan, Neary, Mussa, and Sekiguchi and Horiuchi, among others.[26] Their arguments can be summarized as follows.[27] Figure 4-2 illustrates the adjustment of resource allocation in an economy due to the change of the relative price. In this figure, we assume that X_1 is the importable good and X_2 is the exportable good. Thus, the relative price of X_1 to X_2, which is shown as P_0 and P_1 in the figure, also means the external terms of trade. The curve LL' is the long-run production possibility frontier of an economy. The initial production point is Q_0, where relative price P_0 is tangent to the long-run production possibility frontier LL'. The consumption point is at C_0.

Let us assume that the price of the importable good X_1, like the price of primary aluminum, falls, changing the relative price from P_0 to P_1. The new long-run equilibrium production point is then Q_1 and consumption is C_1. However, capital, unlike putty, cannot be transformed instantaneously without cost from the production of X_1 to the production of X_2. Under this short-run capital specificity, the available transformation curve is SS', not LL'. Thus, at the first stage of the adjustment, the production point is Q_2. At this stage, national income in terms of X_2, that is, OM, is smaller than when adjustment is completed, that is, ON. This differential MN is the adjustment cost associated with the transition of the economy to a new equilibrium. However, as reallocation of capital gradually takes place, the short-run transformation curve shifts up along the long-run transformation curve. In this process the production point gradually approaches Q_1, and the national income in terms of X_2 increases to ON. The adjustment cost that arises during the transition process is socially necessary to attain the new equilibrium at the new relative price. In the real world in which expectations are not always formed with perfect foresight and the social rate of discount differs from the private rate of discount, public policy is needed to encourage the shift of the factors of production from declining industries to expanding industries.

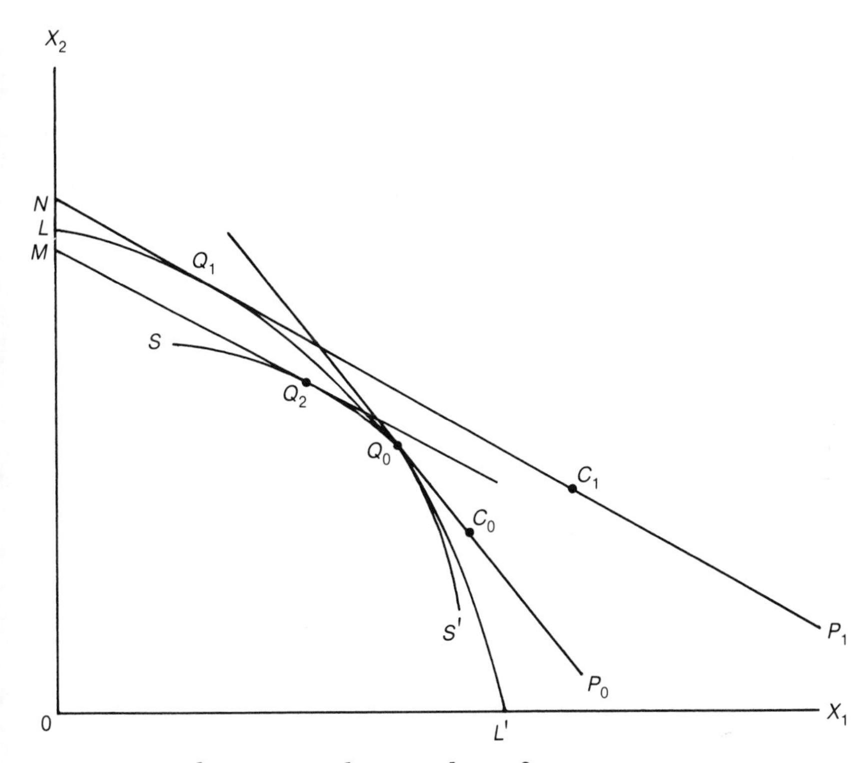

Figure 4-2. Adjustment under capital specificity.

If the wage rate is sticky, in addition to the capital specificity, the labor force that becomes redundant in the declining X_1 industry is not employed by the expanding X_2 industry; as a result a portion of the labor force becomes unemployed. In terms of the figure, the production point moves inside of the short-run transformation curve.

The loss to the economy because of this type of market imperfection can be and should be avoided by public policy. Ideally, public policy should eliminate the imperfections at the factor market to make the short-run transformation curve closer to the long-run transformation curve. Although it is apparent that this is the best policy to implement, it is impossible to eliminate the imperfections at the factor market in a short time. Thus, in the real world, the target of public policy toward declining industries is to carry out the difficult task of encouraging the shift of factors of production from declining industries to expanding industries without massive unemployment.

The Japanese adjustment assistance policy toward the aluminum industry for the most part was successful in carrying out this task. Public policy encouraged the elimination of excess capacity by providing subsidies through the tariff quota and the tariff exemption systems. Efforts were also made to encourage the shift of the labor force. Finally, the adjustment assistance policy encouraged the industry in its attempts to reduce electricity costs as well. With these policy measures, public policy was successful in holding down the loss and pain associated with the drastic reduction of the aluminum industry over such a short period. What was not done was to raise tariff rates and provide subsidies to maintain a certain amount of capacity artificially, in contrast to some of the European countries where governments provided subsidies to keep the aluminum industry intact. Indeed, considering the tendency of government policy in many countries and industries to preserve existing jobs and firms at any cost, this choice of Japanese public policy is remarkable.

What is less apparent is whether this choice reflects the economic philosophy of MITI. MITI did try to maintain a certain amount of domestic capacity, as indicated by the fact that the aluminum industry stabilization plan was based on the assumption of keeping 1.1 million tons of capacity and the structural improvement plan was based on 700,000 tons. The arguments that were developed in these plans for keeping a certain amount of domestic capacity cited "national security," the fear of domination by the international oligopoly, and the instability and insecurity of depending heavily on imports, especially in the case of shortages. Therefore, it is probably correct to assume that the choice of public policy toward the aluminum industry, with its emphasis on assistance for adjustments rather than on keeping a certain amount of capacity, was more the result of factors arising from current conditions than from a philosophy of free trade.

The first of these factors was that the loss of international competitiveness was so severe and so apparent. The subsidies necessary to keep a certain amount of capacity operating, both then and now, would have been enormous. Such subsidies are simply not viable under current governmental fiscal conditions.

The second factor involved the impact of policy on the fabricators. To erect a tariff barrier to protect the smelters would raise primary aluminum prices for fabricators, which are gradually replacing smelters as the central figures in the Japanese aluminum industry. To protect smelters at the expense of fabricators was not politically possible.

The third factor was the declining market power of the big six in the world aluminum market. One of MITI's concerns was the fear of domination of the domestic market by the multinationals of the United States and Europe, a concern that affected the ministry's policy formulation in industries like automobiles, petroleum, computers, and camera film. If the domestic markets of these industries had been dominated by the multinationals without domestic competitors, the argument went, Japanese consumers might have to pay higher prices and "national security" might have been endangered. In the aluminum smelting industry, however, the market power of the big six declined as the developing countries and the Eastern European countries began and then expanded production. In addition, direct overseas investment by Japanese smelters and fabricators increased considerably in the 1970s. Such projects are considered to provide a stable, dependable source of primary aluminum, thus precluding the necessity of a large domestic capacity.

The fourth factor was the relatively small number of workers in this industry. Unlike the coal mining industry, in which the collapse of the industry meant the collapse of several regional economies, the situation in the aluminum industry was not that severe because the number of aluminum workers was relatively small—only 15,000 in 1975. In addition, because the producers belonged to business groups, other members of the groups, as well as the parent companies, helped to absorb the excess labor force from closing or declining smelting firms. About one-half of the employees who left aluminum-producing companies between 1982 and 1984 were hired by the parent and the group companies, a circumstance that made the industry's adjustment much less painful.

A fifth factor was pressure from the U.S. government, which was concerned about the amount of its trade deficit vis-à-vis Japan. The United States voiced concern over the Japanese policy toward the "structurally depressed industries" in general, and while the two laws described earlier were being planned by MITI, the U.S. government warned the ministry not to take protectionist measures that would prevent U.S. firms from exporting more products. With the aluminum industry in particular the United States also demanded the reduction of tariff rates. Finally, the United States objected strongly to MITI's idea of discouraging Japanese imports of primary aluminum on a spot basis by making such imports exempt from the tariff quota system. MITI's intention in this instance was to secure a stable supply of aluminum by encouraging imports on a long-term contract basis and from the offshore projects by Japanese firms. This specific feature was subsequently abandoned, however, and all imports were included in the tariff quota system, partly because of objections by the United States (an exporter of ingots to Japan on both spot and long-term contract bases).

All of these factors contributed to the adoption of the Japanese policy of adjustment assistance. Its main focus was to gain time for the producers so that they could decide how to adjust to the sudden increase in the price of oil and to support basic research on the new smelting process. The subsidies provided through the tariff quota and tariff exemption systems were introduced with a clear limited duration to avoid the creation of vested interests. In addition, public policy did not permit the subsidizing of a certain amount of capacity under the guise of "national security" or for other reasons. Rather, Japanese industrial policy has been to encourage the growth of industries by working with market forces and not replacing the market with the political process. In the end, public policy to assist the adjustment of the aluminum industry in Japan seems to have worked in just this way.

NOTES

*Publications with an * are in Japanese.*

1. For an evaluation of the role of industrial policy in Japan, see R. Komiya, M. Okuno, and K. Suzumura, eds., *Industrial Policy in Japan** (Tokyo University Press, 1984) (English version forthcoming, Academic Press, 1987).
2. Fair Trade Commission, *Industrial Concentration in Japan** (Tokyo, Toyo Keizai Shinposha, 1957).
3. The data in this chapter on the Japanese aluminum industry, including those on production, imports, demand, and prices, are drawn from the Japan Light Metal Association, *Light Metal Statistics in Japan** (Tokyo, various years).
4. According to estimates by the Japanese Ministry of International Trade and Industry (MITI), the demand for primary aluminum will grow by 7.3 percent annually in Japan during the first half of the 1980s; the demand in the other market economies combined will grow by 4 to 5 percent.
5. Kyutai Tanaka, *Current States and Problems of the Japanese Aluminum Industry** (Tokyo, Light Metal Association, 1969). See also Kazunori

Echigo, "Aluminum,"* in H. Kumagai ed., *Industrial Organization in Japan*, vol. #2 (Tokyo, Chuokoron, 1973).

6. Echigo, "Aluminum."
7. Tanaka, "Current States and Problems."*
8. See Richard P. Rumelt, *Strategy, Structure, and Economic Performance* (Cambridge, Mass., Harvard University Press, 1974); and Akira Goto, "Statistical Evidence on the Diversification of Japanese Large Firms," *Journal of Industrial Economics* vol. 29, no. 3 (March 1981) pp. 271–278.
9. Group affiliations of the smelting firms are as follows: Nippon Light Metal—Daiichi Kangin Group; Sumitomo Chemical (the parent company of Sumitomo Aluminum)—Sumitomo Group; Showa Denko (the parent company of Showa Aluminum)—Fuji (Fuyo) Group; Mitsubishi Chemical (the parent company of Ryoka Light Metal)—Mitsubishi Group; Mitsui Aluminum—Mitsui Group; and Sumikei Aluminum—Sumitomo Group. As to business groups, see Akira Goto, "Business Groups in a Market Economy," *European Economic Review* vol. 19, no. 1 (September 1982) pp. 53–70.
10. Japanese yen are converted to U.S. dollars in this instance using the exchange rates then current.
11. Tanaka, "Current States and Problems."*
12. Ibid.; Sangyokozo Chosakai, *Industrial Structure in Japan,** vol. 3 (Tsusho Sangyo Kenkyusha, 1964).
13. The reason this estimate of an 8 percent drop is the highest possible is that not all the materials were imported. Because we could not obtain the ratio of the cost of imported materials to the total cost, we could only estimate the upper limit of the possible cost reduction. See Masaaki Tashita, "International Competitiveness in the Aluminum Smelting Industry and the Investment in Plants and Equipment,"* *Chogin Chosa Geppo* no. 146 (January 1976) pp. 3–31.
14. Ibid.
15. Ibid.
16. Yoshihiro Kimura, "Withdrawal in the Aluminum Smelting Industry and the Future Problems,"* *Chogin Chosa Geppo* no. 202 (March 1983) pp. 3–33.
17. Ibid.
18. As of October 1986, all of the remaining smelters except one were already closed or reported to be in the process of closure. Within a few years, the Kanbara smelter of Nippon Light Metal may be the only smelter in Japan.
19. The volume of production actually fell faster than expected. In 1985, it was 209,000 tons.
20. In 1976 the composition of the electricity of power companies by power source was as follows; hydro, 17.3 percent; coal, 4 percent; oil, 63.1

percent; gas, 8.9 percent; and nuclear, 6.7 percent. See *1977 Energy Statistics* (Agency of Natural Resources and Energy, Tsusho Sangyo Kenkyusha, 1977).

21. The Industrial Structure Council, "The Report of the Committee on Nonferrous Metals, the Industrial Structure Council,"* *Aluminum*, no. 653 (January 1985) pp. 19–24.
22. For the scheme of the Research Association of Mining and Manufacturing Technology, see A. Goto and R. Wakasugi, "Technology Policy," in R. Komiya, M. Okuno, and K. Suzumura, eds., *Industrial Policy in Japan** (Tokyo University Press, 1984) (English version forthcoming, Academic Press, 1987) pp. 159–180.
23. Japan Aluminum Federation, "Current States of Aluminum Industry"* (mimeo, Tokyo, June 1982).
24. For an explanation of these two laws, see MITI, *The Interpretation of the Temporary Measures Law for the Stabilization of Specific Depressed Industries** (Tsusan Chosa Kai, 1978); and MITI, *The Interpretation of the Temporary Measures Law for Structural Adjustment of Specific Industries** (Tsusan Chosa Kai, 1983).
25. After 1985, instead of exempting producers completely from tariffs up to a certain amount (i.e., the capacity they had closed in response to the plan), a 1 percent tariff is levied up to this amount. Beyond it, the 9 percent tariff is levied as before 1985.
26. Wolfgang Mayer, "Short-run and Long-run Equilibrium for a Small Open Economy," *Journal of Political Economy* vol. 82, no. 5 (Sept.–Oct. 1974) pp. 955–967; Harvey E. Lapan, "International Trade, Factor Market Distortions, and the Optimal Dynamic Subsidy," *American Economic Review* vol. 66, no. 3 (June 1976) pp. 335–346; J. Peter Neary, "Intersectoral Capital Mobility, Wage Stickiness, and the Case for Adjustment Assistance," in J. N. Bhagwati, *Import Competition and Response* (Chicago, University of Chicago Press, 1982) pp. 39–71; Michael Mussa, "Government Policy and the Adjustment Process," in J. N. Bhagwati, *Import Competition and Response* (Chicago, University of Chicago Press, 1982) pp. 73–122; and Sueo Sekiguchi and T. Horiuchi, "International Trade and Adjustment Assistance,"* in R. Komiya, M. Okuno, and K. Suzumura, eds., *Industrial Policy in Japan** (Tokyo University Press, 1984) (English version forthcoming, Academic Press, 1987) pp. 327–344.
27. For this point, the author benefited from discussions with Sueo Sekiguchi.

5

AUSTRALIA: ONE DAY IN THE SUNSHINE

JOHN J. BEGGS

As chapter 1 points out the Australian aluminum industry is often grouped with the winners in the shifts of international competitiveness of the 1970s following the two oil shocks. Yet as the title of this chapter suggests, this author at least has certain doubts about whether this is so. An explanation of these doubts requires analyzing the short and what is perhaps the not too happy life of the Australian aluminum industry.

The industry's development began with the intensive activity that followed the discovery of massive deposits of bauxite in the mid-1950s at Weipa in North Queensland, at Gove in the Northern Territory, and in the southwest region of Australia. (The small smelter that had been built earlier in Boll Bay, Tasmania, was a wartime aberration.) Mining of the bauxite began, and exports to the Japanese and European markets expanded rapidly. By the 1970s the increasing scale of the mine at Weipa had made it the world's largest shipping point for bauxite; the dramatic decline in dry bulk shipping costs, through the use of large bulk carriers, made these exports competitive. Concurrently, the country's alumina refining and aluminum smelting industries were also developing. As a result, alumina refining quickly became a major export industry, with Australia accounting for almost 25 percent of alumina output in the market economies by the late 1970s.

John Beggs is in the Department of Statistics and Economics at the Australian National University.

In contrast to its spectacular successes in mining and refining, however, by the late 1970s Australia still had only three small smelters, and their combined capacity was no more than 260,000 tons annually, with approximately 30 percent of production exported. These smelters were developed to supply the domestic market; their financial viability was guaranteed by the presence of protective trade barriers (table 5-1).

Yet the seeds of Australia's grand flirtation with the aluminum smelting industry were being sown in the early 1970s,[1] beginning with the energy price shocks induced by the Organization of Petroleum Exporting Countries (OPEC) in 1973 and 1979. The sudden rise in oil prices affected aluminum producers worldwide. Japan, for example, with its largely oil-based electricity supplies, cut back its smelting operations markedly; Europe and the United States were also affected, although perhaps not as severely, as prices of all thermally generated electricity began to rise. In addition, when the various long-term contracts for the purchase of hydroelectricity expired, the price of this electricity would rise still higher. These circumstances—combined with delays in the development of nuclear power and the surge in demand for aluminum and the resulting high spot prices in 1979–1980—encouraged the search for other sources of competitively priced electricity.

Attention soon focused on Australia's attractive combination of abundant raw materials and low-cost, coal-based electricity, and six reasonably firm proposals were made to build smelters at Boyne Island, Portland, Tomago, Lochinvar, Bundaberg, and Westal. In addition, expansions were planned at the existing smelters at Kurri Kurri, Point Henry, and Bell Bay. Together, these new smelter projects were to constitute about one-third of the Western world's new

Table 5-1. Tariff Rates on Primary Aluminum, 1955–1986 (percentage)

Time period	General tariff
January 1, 1955–May 13, 1964	12.5
May 14, 1964–July 18, 1973	7.5
July 19, 1973–June 30, 1974	5.625
July 1, 1974–August 10, 1976	6
August 11, 1976–December 31, 1976	6, and 1.5 for each cent by which FOB price is less than $0.71/kg
January 1, 1977–December 31, 1982	6
January 1, 1983–1986	2

Source: Data obtained by the author from the Australian Department of Trade.

capacity proposed by the industry in the 1980s. A similar proportion was planned for South America, mainly in Brazil. Only 20 percent of the planned new capacity was to be located in the older areas of production, and the majority of that was slated for Canada.

The actual growth of production capacity in Australia in the late 1970s and early 1980s has been substantial, and annual capacity is now approximately 800,000 tons. Two of the proposed new smelters (Boyne Island and Tomago) have been built; a third (Portland) has recommenced construction after a delay caused by uncertainty over world aluminum prices. In addition, the planned expansions have been completed at Kurri Kurri, Point Henry, and Bell Bay. Yet, the full extent of the "boom" prospects of 1980 has not been realized, and the past four years have seen a wide-ranging public debate in Australia on both the desirability and viability of this industry in the nation's economy. The point of departure for this chapter is the domestic public policy issues and central themes in this debate, which extend over matters of energy policy and the pricing of electricity, the implicit pricing of coal reserves, foreign ownership and control of domestic industry, the operation of wage negotiations and labor relations, macroeconomic employment effects, taxation, the environment, and aboriginal land rights. The lessons to be learned from the experience are many and will remain useful for many years, both to those involved in the formulation of Australian policy and to policymakers and economists in other countries.

ELECTRICITY: THE HEART OF THE MATTER

Australian commercial electricity is supplied almost exclusively by authorities owned and controlled by the state governments. This point is central to understanding a great deal of the controversy surrounding the development of the aluminum smelting industry in Australia. Most of the incremental electricity for industrial use comes from coal-fired generating stations; the electricity authorities hold proprietary or other rights over the coal reserves needed to supply the generators. Furthermore, the rail systems used to haul the coal are owned by the state governments. Consequently, very few "arm's-length" transactions take place between the mining of coal and the point-of-sale of electricity to the final user.[2] From this amalgam of government ownership and regulation arises the central question: What price should be charged for electricity?

Some legal and historical perspectives are essential. The Australian constitution creates a commonwealth of states in which residual powers, that is, those not explicitly assigned to the central government in the constitution, remain with the individual state governments. Each state government has insisted strongly on its legal rights, and the intense expression of these rights has also manifested itself as interstate commercial rivalry. In the case of major development projects, state governments may vie with each other to offer scarce resources to developers on the most favorable terms—favorable to the developer, that is. Such practices are justified because growth and economic development are seen as paramount virtues (especially at election time), even when purchased at such high costs as to provide very little benefit to the communities concerned.

The extent of this destructive interstate rivalry is unclear, but it should be stressed that the federal government has no legislative power to restrict it. The Australian government cannot direct the behavior of the states, cannot gain access to data used by the states in policy formulation, and cannot even acquire data on contracts entered into by the states. (For example, the federal government cannot find out the price that states charge the aluminum smelters for their electricity.) The only roles left to the federal government are to run public enquiries that expose the issues associated with major development projects[3] or to frustrate state government by the selective use of laws relating to such matters as the environment and foreign trade. Both courses of action are typically perceived as interference in the sovereign affairs of individual states of the federation and are usually judged to be too politically unpalatable to pursue.

Although the question of the prerogatives of state governments versus those of the federal government is central, it is important to recognize the limited legal ability of citizens in Australia to gain access to documents relating to internal decision making at the state government level. The federal government has passed reasonably comprehensive "freedom of information" legislation, yet the power of this legislation does not extend to the operation of state governments. Some state governments have passed similar legislation, but in general it does not extend to the operation of state government enterprises such as coal mines, railways, and electricity authorities. This barrier to public scrutiny tends to protect an entrenched bureaucracy dominated by engineers who often have little training in economics and who hold an implicit preference for policies that favor growth. In such circumstances, pricing policies and cost allocation procedures develop that are more likely to suit these narrow bureaucratic and political goals rather than the efficient allocation of scarce resources.

All of these issues find a sharp focus in the debate over the pricing of electricity to new aluminum smelters.[4] The picture for electricity authorities as it appeared in 1980 for the coming decade was one of significant expansion of existing coal-fired generating capacity, with the largest share of this growth going to one industry—aluminum. As the decade has progressed, three particular issues stand out in this debate and are addressed in the following pages. First, if there is any truth in the saying that aluminum is "solid electricity," in the Australian context, it can also be termed "congealed coal" because it is coal that generates the electricity. The pricing of coal thus plays a major role in determining the viability of the industry, both from the perspective of the aluminum companies and from the view of the general community. Second, there is the issue of the special characteristics of the electricity load pattern of aluminum smelters and the contractual arrangements that have been made to accommodate this factor. Third, and perhaps most important, is the way in which electricity authorities allocate costs and determine their required rates of return on investment in new power-generating facilities. Errors in this area can lead to profound long-term misallocations of resources (the Loy Yang power station to supply the Portland smelter being a rather celebrated case in point).

Coal Pricing

The pricing of coal consumed by state electricity authorities has been a major subject of controversy. Critics claim that the underpricing of coal compared to its shadow price (or market value) is a major component of the electricity "subsidy" going to aluminum smelters—that is, a subsidy in the form of income forgone by the community because of the failure to obtain the best price for coal.

To add substance to the discussion, it is useful to consider the Australian Department of Treasury definition of the correct price for coal. This price is termed the export parity price:

> For purposes of valuing coal used in electricity generation, the export parity price of steaming coal can be defined as the price at which the next additional tonne could be exported, excluding freight, insurance and hauling charges. To the extent that a private producer regarded a long-term contract with a domestic electricity authority as less risky than production for export (where, of course, long-term contracts are commonplace also), this lesser risk would need to be taken into account

in determining the appropriate price to be charged to the domestic authority. To the extent that some or all the coal used by Australian electricity authorities is a different quality from that exported, that quality difference would also need to be reflected in the domestic price.[5]

Separate debates have centered on Australia's three main sources of coal for electricity generation: (1) brown coal from Victoria, (2) underground black coal from New South Wales, and (3) open-cut black coal from Queensland. The overriding principles are the same for all the sources, but each case provides insights into the particular dilemmas faced in assigning a meaningful opportunity cost to coal reserves. Given the vast reserves of coal in Australia, it is difficult to compute the scarcity value of these resources. Only 2 to 3 percent of these coal reserves will be exhausted for electricity production over the next fifty years, even after making allowances for relatively high rates of growth in electricity consumption and full implementation of the proposed aluminum smelting projects (table 5-2). The argument follows that extraction costs fully reflect opportunity cost, particularly in those grades of coal for which there is no current export market.

Individual states value their coal reserves differently. The Department of Treasury has estimated that the unit values placed on coal per kilowatt-hour of electricity generated by the utilities in 1978–1979 were 0.306 cents in Victoria, 0.290 cents in New South Wales, 0.546 cents in Queensland, and 0.870 cents in Western Australia (all figures are in current U.S. cents).[6] These differences are substantial.

Table 5-2. Depletion Rates of Coal Reserves: Total Usage Over Fifty Years (absolute numbers in millions of tons)

		Electricity[b]		Smelters only[c]	
State	Reserves[a]	Quantity	Percentage	Quantity	Percentage
New South Wales	182,100	2,878	1.6	198	0.1
Queensland	71,800	1,692	2.4	288	0.4
Victoria	107,000	6,023	5.6	104	0.6

Source: Senate Standing Committee on Natural Resources, *Report on the Development of the Bauxite, Alumina, and Aluminum Industries* (Canberra, Australian Government Printing Service, 1981) p. 15.

[a] Demonstrated and inferred.

[b] Based on 3 percent per capita growth rate in consumption per year.

[c] All aluminum smelter projects proposed in 1980 come on line.

First, let us consider Victorian brown coal. With its low thermal content, there is no export market for brown coal, and the presence of high-grade coal deposits in other states removes the possibility of domestic marketing. The competitive alternative uses of brown coal are limited to local electricity generation and possible local synfuel production in the future. Yet because electricity will continue to be produced by existing state government authorities and synfuels remain an unmeasured quantity, the market for alternative uses provides little information about the present scarcity value of the resource. Simple cross-state comparisons based on the thermal content of the coal, such as those cited above, can be used to indicate scarcity value, but the matter is confused by differences in handling costs for different grades of coal and also by the more capital-intensive generating stations required for brown coal.

The situation in New South Wales is the most contentious. The state electricity authority has never accepted the principle of opportunity cost pricing, which in the case of black steaming coal is equivalent to export parity pricing. In its costing procedures the state authority charges not what the coal is worth on the open market but the "out-of-pocket" cost of mining and delivery to the power station. When it has been questioned about this practice, the authority's response has been that the steaming coal used in power stations is of such low quality that it is not exportable and therefore of no alternative use.

There is some prima facie evidence for this position. Almost all coal used by the electricity authority has an ash content in excess of 20 percent, which is about the maximum ash content possible for exportable coal. Nevertheless, high-ash coal is not necessarily cheap to use. The higher the ash content, the greater the tonnage of coal that must be mined and transported to generate a given amount of heat. (For example, burning coal with an ash content of 30 percent requires about 20 percent more coal to produce the same amount of electricity as coal with an ash content of 15 percent.) The unit cost per ton is lower, but savings in this area are offset by the greater quantity consumed. Boiler tube abrasion also is more severe in high-ash coal. Boilers can be designed to burn such coal, but it is still the case that maintenance costs are higher and boiler life shorter. In addition, wear and tear on the milling plant and ash precipitators are much greater with high-ash coal.

Thus, savings in fuel costs are offset by higher maintenance and capital costs. H. W. Dick has suggested that these costs have been significantly understated in past decision making by the electricity authority.[7] In 1981 the authority announced a large investment program to install washing plants at many mines supplying generating stations. The costs of washing are about $3.50 per ton, and about 25 percent of the coal is lost in the process.[8] Once the coal is washed, however, it becomes of export quality, thereby completely undermining the argument that it has no other use and should be valued at the cost of production.

Queensland steaming coal is sold to the electricity authority under a number of different circumstances. Most coal is sold at prices established by the Queensland Coal Board, a statutory marketing authority that sets prices that generally correspond to the notion of export parity pricing. Some coal, however, is supplied to the generating authority at cost under special provisions written into the contracts the state government issues to develop a mine site. A case in point is the coal provided to the Gladstone generating station from the Utah-Blackwater mine. For the mining developer, the costs of this coal are seen as an implicit resource rent tax that the state is levying for access to the coal reserves.[9] The only problem arises when purchases of the coal are entered into the books of the electricity generating authority. Clearly, this coal should be assigned its true opportunity cost (or export parity price) so that the ultimate users of electricity bear the true scarcity cost of the resource consumed in generating electricity. Accounting convention, however, leads this coal to be priced at only the direct (extraction) costs to the generating authority, with a subsequent downward distortion of electricity prices.

Electricity Load Characteristics

The demand for electricity by aluminum smelters is subject to minimal day-to-day or seasonal fluctuations. This stable quality contrasts sharply with the characteristics of overall demand on electricity grids in which peak load may be as much as twice the minimum load. Generating authorities minimize average costs by meeting a high proportion of their maximum load from base-load stations, which run twenty-four hours a day so that the system can provide a virtually instantaneous response to increased demand. Base-load customers allow these stations to operate economically around the clock by smoothing out the peaks and troughs of the pattern of demand for power. An additional advantage of aluminum smelters as electricity consumers is that the smelters' supply can be interrupted for periods of from two to three hours, enabling suppliers to divert power to peak demand elsewhere. This interruptibility is recognized in all power contracts made with Australian smelters. Although interruptibility is often seen as a special virtue of smelters as electricity consumers, there are many questions surrounding the assessment of the benefits of this virtue because of the type of interruptibility permitted by the contracts.

It is a fact that electricity generation is not totally reliable, and it matters little whether the causes of the unreliability are technical problems or the result of poor industrial relations. The nature of smelters is such that power supplies cannot be interrupted for more than three hours or molten aluminum will freeze in the smelting pots, causing damage that may require shutting down the plant for several months. The resulting cost to the smelting companies of these shutdowns is enormous. Consequently, although short-term interruptibility can be viewed as an advantage of smelter demand for the supply utility, long-term vulnerability is likewise a disadvantage. When electricity production is restricted for more than three hours, smelters must continue to draw their full load requirement almost irrespective of other needs in the system, and there is concern on the part of electricity authorities that existing contracts with aluminum smelters do not address this risk adequately. Although clauses of contracts signed with aluminum companies allow immediate cutoffs of up to three hours, the frequency of these cutoffs is limited to no more than once per fortnight.[10] Until the cutoff right is exercised, a smelter is interruptible and can be thought of as a spinning reserve. But it is a spinning reserve with a three-hour limit; once that reserve is used, it becomes unavailable for two weeks.

This undesirable situation is the direct result of injudiciously written contracts between the generating authority and the aluminum companies. As press disclosures have heightened public awareness of the features of these contractual arrangements, community unease has increased. A policy of fully revealing the public interest implications of such contracts would improve community attitudes toward this kind of development.

The interruptibility issue is more important in Australia than it is in some other countries—those that supply smelters from hydro generating systems. Australia uses thermal generating systems, which have certain characteristics that make them more prone to trouble than hydro systems. For one thing, the physical complexity of a coal-fired plant is greater than that of a hydro station, and there is the additional need to maintain a supply of coal. A particular problem appears to be the reliability of the large-scale, coal-fired generating stations (660-MW units are the current standard) that are being built to service the new demand created by the expansion of the aluminum industry. The technology of large, base-load units is being imported to Australia from Europe and North America, and the transposition of the technology is not going smoothly, at least in the early stages. In many of the large new units the heat transfer balance between furnace tubes and superheater-reheaters is different from that given in the manufacturer's specifications (the difference is probably due to the characteristics of local coal, which can be quite different from typical North American and European coals). More distressing, however, is the fact that many of the potential economies of scale of large-scale generators are not being realized. The idle time that results from faults in large generators is proportionately larger, and new systems take longer to settle down to a period of reasonably trouble-free operation. The greater downtime in large units seems to reflect their greater complexity and the fact that the length of internal tubing increases more or less proportionately with size.

Nevertheless, there is every reason to believe that recent technical problems with the reliability of large-scale base-load units can be overcome through a process that involves the industry moving down an implicit learning curve of the sort found in many other industries. The central point is that this learning is necessarily a time-consuming process, which must, in the short run, limit the economically viable rate of expansion of the industry, although large-scale operations are still likely to be viable in the long run. The entire process, however, has implications for the development of the aluminum industry. On the one hand, very rapid development of aluminum smelting will be associated with rapid and less efficient expansion of the nation's electricity generating capacity. If, on the other hand, the development of the smelting industry proceeds more slowly, there will be enough time to learn the most efficient methods of generating electricity in large coal-fired units under local conditions.

Capital Costs of Power Generation

The price to be paid for electricity by aluminum smelters has been the single most controversial question arising out of the whole recent development of the industry. The controversy has arisen as a result of a widespread belief that the community at large is subsidizing the provision of electricity to smelters. This is argued to occur either directly, as a result of prices that are set at less than the cost of production, or indirectly, in the form of income forgone by underpricing the various factors of production (see the coal example discussed earlier). We have already discussed some particular problems in valuing Australian steaming coal. The difficulties of assigning the correct opportunity cost to capital are more general issues that belong in the category of public economics, but they have taken on special significance in the Australian public debate of the aluminum industry. Most of the discussion has focused on two issues: first, the treatment of risk or uncertainty in establishing the required rate of return for electricity generating authorities; and second, the cost allocation methods used to establish the asset base on which a required rate of return should be earned. (This situation is quite different from those found in Canada and Brazil as far as the pricing of electricity to the aluminum industry is concerned. In those countries, electricity to the industry is supplied from large, already completed—or, at the least, committed—hydropower projects.) In the sense that "bygones are bygones" from an opportunity cost perspective, the pricing problem is simply to charge an electricity price that maximizes revenues. In Australia, the power stations to supply new smelters remain to be built. The opportunity cost of the resources that will be committed to these projects should be correctly assessed.

In assessing the required rate of return of electricity authorities, we should look at whether the nominal pretax rate of return on the facilities equals its cost of capital, based on the weighted average of debt and equity capital and including an allowance for risk and the expected rate of inflation. The imputed allowance for risk should be the same as the allowance that would be provided if the electricity generating capacity were in the hands of a private investor. (An example would be Alcoa's Anglesea plant, which provides a portion of the power for its Port Henry smelter.)

Although electricity authorities raise funds by the public sale of bonds, this interest rate is generally not an appropriate one for assessing the cost of capital for power stations. Typically, these bond offerings are guaranteed by the state government; as a result, they understate the true nominal interest rate because they represent the average riskiness of all government loan raising rather than the marginal contribution to risk of, for example, a power station. Bond-financed capital is the debt capital of the electricity authority, and the residual ownership of the authority is held by the taxpayer who becomes an involuntary risk holder and hence an implicit equity shareholder. When an organization such as a state electricity authority gets into financial difficulties, it is the taxpayer who ultimately has to meet the financial commitment on borrowings. Just as the equity investor in the firm needs to be compensated for voluntarily assuming this risk, so should the taxpayer—who has far less choice in the matter.

Available evidence suggests that the choice of required rates of return by state electricity authorities in Australia bears little relation to the type of economic intuition developed in the preceding paragraph. P. L. Swan reports:

> The Southern Electricity Commission of Victoria (SECV) used an 8.5 percent real discount rate for evaluating Loy Yang (the power station to supply Alcoa's Portland smelter) although it did not specify at the time that it was in real terms. However, the SECV to my knowledge has not advocated the use of the same discount rate it uses for project evaluation and costing purposes for tariff setting and pricing purposes. The implicit rate used for tariff setting purposes is probably negative since until the 1979–80 Annual Report the SECV set itself the objective of earning 7 percent in nominal terms (about minus 3 percent in real terms) on its funds. Nowhere to my knowledge has the SECV provided a public justification for the use of an 8.5 percent real rate for one purpose and about minus 3 percent for another when logic dictates that consistency is required.[11]

SECV is unique among the Australian electricity authorities in providing at least some public accounting of the decision criteria it employs. Establishing sensible, required rates of return for large public projects is a clear priority when evaluating the economic desirability of such new developments as an aluminum smelting industry.

The Australian Department of Treasury has argued that the appropriate real rate of return to be used for all public projects is 10 percent, with a sensitivity range of from 7 to 13 percent.[12] This is an overview type of measure based on comparisons with the private sector generally and does not account for the idiosyncratic risk of a particular sector or even of a particular large project. P. L. Swan, however, reports a series of computations that attempt to allow for the riskiness of electricity investment itself. Because Australian electricity production is a monopoly in the hands of state governments, there are no stock market data available to compute the appropriate, real risk-adjusted cost of capital. In the United States, however, privately owned electricity utilities are traded on U.S.

stock exchanges; thus, it is possible to compute a risk premium for the U.S. industry and, as a rough comparison, to extrapolate this number to Australian data. Using a capital asset pricing model approach (which involves computing a measure of the systematic or nondiversifiable risk associated with electricity utilities[13]), Swan concludes that state electricity authorities "should be encouraged to aim for a rate-of-return on capital of around 8 percent in real terms. . . . which would appear to represent approximately the required return if the enterprise was in private hands."[14] This proposal is in sharp contrast to current SECV publicly announced policy, which represents a rate of return of minus 3 percent in real terms.

Many serious anomalies exist in the cost allocation practices of state authorities—in particular, the electricity authorities. SECV is a case in point. In its costing of the Loy Yang power station, which was built to supply Alcoa's Portland smelter, SECV did not allow for interest on expenditures incurred during the sixteen or more years of the construction phase of the power station. For instance, in 1967 SECV reported that "interest during construction will be met from the Commission's operating accounts."[15] Hence, these costs were neither capitalized as part of the basic asset on which the project must earn a required rate of return, nor were they treated as an operating charge against the Loy Yang project. Excluding interest during construction leads to gross understatement of capital costs; it also has the effect of biasing project selection toward large projects with long construction periods and against small projects with rapid payoffs. The details of this issue are too complex to canvas at length in this chapter,[16] but it does seem that biases of this type favor the kind of power facility required for aluminum smelters and hence create an artificial preference for an industry that would be judged economically inefficient if the costs were properly assessed.

In 1982, in the face of proposed increases in electricity tariffs, Alcoa decided not to proceed with construction of its Portland smelter. Aside from the political embarrassment, this decision left the Victoria government with an electricity transmission system to Portland that had no alternative use and a partially constructed power station whose alternative uses were limited in the short term. The state government has recommenced negotiations with Alcoa on electricity tariffs and other matters in the hope of resolving this situation.

Of particular concern is the direct construction cost of the powerhouse at Loy Yang. As will be discussed later in this chapter, both the direct and indirect costs of labor at the Loy Yang site have risen sharply to levels well above those found for comparable work outside the industry. Consequently, within the Victoria government the argument has been put that the opportunity cost of labor on the site should be measured by the wage that labor would earn off the site rather than by the wage actually paid. (The higher wages reflect the power of the union, thus distorting the true economic cost.) By invoking this reasoning, it is possible to assign a significantly lower capital cost to power generation, which would justify a lower electricity tariff.

The opportunity cost principles espoused in this argument would be disastrous in situations in which wages are negotiated annually. Clearly, when this reasoning is used, the power station construction project would continue regardless of how high wage levels went on the site. Electricity tariffs would be similarly unaffected. The only effect would be a transfer from the public purse to labor employed on the site. The ramifications of this argument, if extended to all public construction projects, would be that everyone would ultimately want to be employed on such projects and the government eventually would be bankrupted. One can only hope, from a distance, that such a mistake will not be made in the final government benefit–cost analysis of Loy Yang.

FOREIGN OWNERSHIP AND CONTROL

It is generally accepted that foreign capital plays an important role in Australia's resource development, and this is particularly true of the aluminum industry. The degree of foreign ownership and control of the industry is relatively great, mirroring the high degree of concentration and vertical integration of production by the industry's major international firms. Foreign ownership in the Australian aluminum industry takes the form of equity in mines, refineries, and smelters and often extends to shares in physical outputs and physical reserves. These practices give shareholders and owners of production rights a certain share of the output of a mine or refinery based on the extent of ownership. Foreign investment has also been accompanied by other specific benefits such as access to current technology, managerial skills, and markets.

The federal government requires that proposed new investments by foreign companies must obtain approval under the government's foreign investment policy, which stipulates that, wherever practicable, Australians should hold some share in major projects. There is no specific minimum equity requirement for new projects involving aluminum smelters,[17] but the project must demonstrate that it will provide significant advantages to the local economy or involve significant Australian ownership before approval is granted.

Foreign investment controls have been motivated by general political concerns together with some quite specific economic concerns. Political sensitivity to nationalist sentiments has led each of the major political parties to articulate and practice policies that limit the intrusion of foreign nationals and foreign corporations into major positions of power in the Australian economy. Most of the concern in the public mind seems to have centered on the agricultural and mining industries. There has been demonstrably less public concern about foreign intrusion in the manufacturing sector. Yet the federal government has also pushed for substantial, independent Australian participation as a means of encouraging arm's-length transactions (particularly in industries in which extractive and processing functions are closely integrated) for the purpose of enforcing taxation law and other financial law.

The taxation of income earned by foreign investors represents a significant part of the overall return to Australia for the resources used by the foreign investor. The company tax rate of 46 percent applies to the taxable income of all companies in the aluminum industry, whether they are locally or foreign owned. In addition to the company tax, dividends declared by resident Australian companies and payable to nonresident shareholders are subject to a dividend withholding tax at the rate of 30 percent; this rate is reduced to 15 percent, however, in situations in which there are comprehensive double tax agreements—as there are with the United States, Canada, France, and Switzerland, the nations in which the big six producers are headquartered. Interest payments made to a nonresident company are also subject to a withholding tax of 10 percent of the amount of interest paid. It is important that these taxes be collected effectively and with a minimum amount of tax avoidance, a feat that requires vigilance on the part of the taxation authorities. Vigilance is particularly necessary with such practices as thin capitalization and transfer pricing, which become difficult to determine without arm's-length transactions. These issues are especially pertinent in the aluminum industry because of the high levels of market concentration worldwide and the extensive vertical integration of the industry.

The problem of thin capitalization arises because of the differential taxation treatment of interest payments and dividends, and the incentive this creates to distort the debt–equity ratio. Interest paid on a debt is a deductible business expense; when it is paid to a nonresident, the tax payable in Australia is confined to a withholding tax of 10 percent of the amount of the interest paid. These arrangements mean the maximum tax payable on $1 of profits (assuming a dividend withholding tax of 15 percent and total distribution of profits as dividends) is 61¢, whereas the maximum tax on $1 paid as interest is only 10¢. The overall tax differences between debt and equity may not be as sharp as this because home country taxes may be reduced by taxes already paid in Australia. Yet the point remains: differing consequences of debt and equity capitalization can create incentives for foreign companies to be thinly capitalized—to the detriment of Australian tax revenue.

The incentive to pay interest rather than dividends usually arises in cases in which loan capital is provided by associated parties in the same corporate group (rather than in situations involving genuine arm's-length transactions). To counteract this incentive in circumstances in which borrowing is not at arm's length, it is often a condition of foreign investment approval that the Australian subsidiary maintain a debt-to-equity ratio of no more than three to one. In the past, aluminum producers operating in Australia have borrowed considerable amounts of capital from arm's-length lenders. Even an apparent arm's-length transaction, however, might involve thin capitalization if the subsidiary in Australia borrows on the guarantee of the subsidiary's foreign parent. In this situation the arm's-length transaction is really between the lender and the foreign guarantor, who effectively then lends to the subsidiary. This type of loan arrangement is therefore taken into account for purposes of the debt–equity ratio test.

The aluminum industry is highly integrated throughout the world. In the situation described above, there is a risk that transfer prices within associated companies will be manipulated to avoid Australian taxation. To counteract such practices, the commissioner of taxation is given the authority to take action whenever a foreign-controlled business produces less taxable income than might be expected if the business were operating using arm's-length transactions. Strong powers[18] are given to the Taxation Office in these situations, although difficulties are sometimes encountered in applying them, particularly given the scarcity of data on which to establish an arm's-length price (particularly for alumina).

The problems of transfer pricing can be ameliorated if the number of independent companies or consortiums operating at each stage in the production chain is increased. Significant local equity in new projects can also help to check practices that are contrary to Australian interests. In addition, from a taxation policy perspective, the increased integration of all stages of production in Australia, in contrast to the export of raw materials to overseas affiliates, helps mitigate the effects of transfer pricing. It is likely that the rather strict foreign ownership controls in the agricultural and mining sectors reflect a somewhat naive public concern that these resources are being sold off overseas at less than their true worth; in situations in which transfer pricing becomes possible, however, these fears seem well founded.

There are also other institutional features of the aluminum industry that are of particular concern in these matters. Most companies in the industry operate under arrangements whereby a subsidiary of a foreign group processes minerals in Australia through a cost-toll company. The cost-toll company makes no profit for taxation purposes and merely carries out joint activities at cost for the companies associated in the business venture. Ownership of the processed minerals normally remains with the participants, and the tolling company charges a fee to cover processing costs, interest, depreciation, and other tax-deductible amounts. The subsidiary typically sells its share of the processed product to its nonresident shareholders or affiliates as inputs to their operations. Under such arrangements, there is obvious scope to undervalue minerals sold to nonresident shareholders. An appropriate tax could prevent the loss of revenues, but it would require special provisions in double-taxation agreements and can only be effected in cases in which such double-taxation agreements exist.[19]

STRUCTURAL ADJUSTMENT

The proposed major expansions of the Australian aluminum smelting industry came during a period in which the nation was entering a broad-based resource boom. Although the boom augured well for a period of relative national prosperity, some analysts became concerned about the effect elsewhere in the economy of the rapid expansion of the mining and mineral-processing industries. These ruminations found expression in what is known in Australia as the "Gregory thesis," or elsewhere in the world as "Dutch blight."[20] The argument runs as follows: if a country is placed in a situation in which it has very rapid and extensive growth of exports from a new sector, then a number of things must follow.

The first is that the more successful the new sector is at exporting a commodity, the more difficult life will be for those other industries that also export and compete with that sector. In the particular case of the Australian economy, the rapid growth of mineral product exports may result in difficulties for those who export agricultural products. Second, the more successful Australia is in exporting in general, the more likely it is that imports will increase, placing pressure on those industries that compete with imported goods. The reason these sectors affect one another is that they are linked together through exchange rates (changes in the rates can alter the profitability of sectors) or through differential rates of inflation in Australia and overseas. If there were rapid growth in the volume of mineral exports, and if the Australian price level did not change quickly, then the exchange rate would appreciate. The appreciation of the exchange rate would make the export of agricultural products less attractive, which would make imports cheaper, and so on. If the exchange rate were "pegged," the same sort of effects would work through the Australian inflation rate.

The popular conception in the financial press regarding the Gregory thesis is that resource development is the harbinger of increased unemployment in those industries that must compete with imports and in existing export-oriented industries. As in the debate over the effect of North Sea gas and oil on the Norwegian and British economies, it appears that a mild but pervasive social antagonism toward resource-oriented industries has found a place in popular thought.

The effects described in the Gregory model may be less stark than they at first seem. The model envisages that the rise in income from the increased value of mineral exports spills primarily into the nontraded goods sector, thereby increasing prices in nontraded goods and attracting labor and capital out of other sectors. If much of the new spending is automatically directed at imported goods or at assets or purchases from the existing export sector—such as would be associated with new aluminum smelters that import capital equipment and purchase export coal in the form of electricity—then there is a far smaller change in the real exchange rate and a much smaller need for structural adjustment. Also, to the extent that investors perceive the mineral boom as transitory and respond by increasing savings, perhaps in the form of capital outflows, the adjustment of real exchange rates will be less pronounced. The recent 1984 liberalization of the Australian foreign exchange market and the clean float of the currency are policy measures that allow the adjustment process to proceed unimpeded without creating further distortionary inefficiency as the ultimate new equilibrium is achieved.

More tangible aspects of structural adjustment are those changes taking place in the labor market. Australian blue-collar workers are extensively unionized, particularly in the construction and transport industries. Furthermore, the unions are organized on a trade or craft basis rather than on an industrial basis. Wage determination is by a process of negotiation between employer and union, followed by government-supplied intermediation or arbitration—or both—on an as-needed basis. Wage settlements are ratified by an industrial commission or court and then become binding on all parties involved. A further complication is created by the existence of separate state and federal wage-determining processes, the former pertaining to employees of firms operating enterprises within the boundaries of a single state. The system is cumbersome in its operation and has significantly impeded industrial development. The effects are quite visible in the recent history of the aluminum industry.

Perhaps the most outstanding example has been the construction of the Loy Yang power station, which is intended to supply electricity to the Alcoa smelter at Portland, Victoria. This project is now several times over its budgeted cost and many years behind schedule. A large contributory factor has been the poor industrial relations at the construction site. Because unions are organized on a trade basis, there was an almost inevitable round of demarcation disputes at the outset to distribute the available jobs among the respective unions seeking to have their members employed on the project. It is difficult to see that any benefits have flowed from these strikes at Loy Yang, other than that of maximizing the power and influence of a small group of union leaders. In fact, the Loy Yang experience has cast doubt over the possible extent and speed of any new developments in the industry.

The policy lesson of this experience is clear. Governments must take a more aggressive position in deregistering unions that behave in a belligerent fashion to the detriment of fellow workers; this sanction should be applied in particular against unions that are repeatedly involved in demarcation disputes.[21] A longer term goal must be the reorganization of unions on an industrial basis to internalize the negotiating process within them and avoid such a burden on the growth section of the economy.

One of the overriding concerns in industrial development is to anticipate the way in which the economic rents from new projects are likely to be shared among participants. The three parties to the process are the firm, the government, and labor. The firm, which is the primary decision maker in project developments, is also the residual legatee for both profits and losses. The government's claims are frequently well known in advance and take the form of a set of resource, income, and sales taxes. Usually, although not always,[22] government taxes and charges can be anticipated well in advance and are not a significant source of uncertainty.

The position of labor is quite different. Labor often follows the strategy of being amenable and inexpensive in the initial stages of a development project. As development proceeds and the firm becomes heavily committed financially, organized labor has an opportunity to extract monopoly rents in the form of wage and conditions settlements. In many cases the long-term economic consequences of these practices have been severe; risk-averse firms have imputed a high probability to the chance of being "wage-gouged" by organized labor and have kept away from otherwise desirable investment opportunities. There can be no doubt that this has contributed to the suspension of proposed aluminum developments in Australia.

The phenomenon of wage-gouging is not new and is certainly not unique to Australia. What is specific to the Australian situation, however, is that such practices were extensive in 1981 and 1982, and in these critical years, many of the decisions on aluminum smelters were being made. The problems stemmed from a conscious decision by the government to move away from the regulated system of wage determination (which had involved hearings and arguments before an industrial commission or court) to a system of collective bargaining that until that time had been a rare feature of Australian industrial life. In principle the intention was to remove some of the institutional rigidities in the wage determination process and give market forces a free reign in setting wages.

The immediate consequences of this policy were disastrous, however. There was an enormous increase in the number of workhours lost due to industrial disputes, and unions in certain key sectors managed to gain very large wage settlements.[23] Those unions whose memberships were mainly involved in large project developments were in an ideal and relatively unhampered position to wage-gouge the developers, and they did so with a high degree of success. Almost all new construction projects, including the aluminum smelters at Lochinvar, Bundaberg, Westal, and Portland, were either canceled or delayed. It is not possible to attribute all of the slowdown in construction to industrial wage-gouging because the world price of aluminum also slumped severely at this time. The low price of aluminum, however, primarily reflected a worldwide recession that was expected to be short lived, and because aluminum smelters and the associated power generating facilities take many years to complete, the slump in itself should not have led to such drastic cutbacks in development.

Since early 1983 the Australian government has operated under an "accord" with the unions that limits wage increases to adjusted cost-of-living increments only. The effects of the accord have been to reduce significantly the extent of industrial disputation and also to stop the rapid growth of real wages that was occurring during 1981–1982. Other, more immediate effects included a rise in business investment and a general recovery of the economy. Many long-term issues remain to be settled, however; holding wages to cost-of-living increases ignores the dictum that changing economic circumstances bring changing wage relativities, both among job skills and among geographic locations.[24] Wage settlement procedures that inhibit this process also inhibit the long-term reallocation of resources and hence economic growth. It remains for the government to wean the unions from the accord in such a way as to avoid another round of wage-gouging with its attendant ill effects on the entire economy.

ENVIRONMENTAL ISSUES AND ABORIGINAL AFFAIRS

As in many countries, major development projects in Australia must supply environmental impact statements to various levels of government before approval is granted to proceed. For the most part, this approval process has not proved to be a significant impediment to industrial developments in the country. The Franklin Dam project in Tasmania proved to be the exception, however, and the decisions in that case have long-term implications for the future development of the aluminum industry in Australia.

The dam on the Franklin River was planned with the potential to supply a large amount of low-cost hydroelectricity to the state of Tasmania. At the time construction began the power was uncommitted, but it was envisaged that the electricity would be sold primarily to new large industrial users such as aluminum smelters. The dam itself was extremely controversial because it was to flood a deep valley that was particularly rich in unique species of flora. The project was eventually abandoned after the Australian High Court in 1983 ruled with the federal government against the state of Tasmania, arguing that under international treaties the central government had the power to stop construction if, in its opinion, there was significant risk of environmental harm. The High Court ruling in the case has been, perhaps, the most controversial in Australia's short history. Its effect was to inject a significant amount of additional federal control over the environmental aspects of project developments, which until 1983 had been the concern of municipal- and state-level governments only.

In another arena of controversy over industrial development—the protection of aboriginal sacred sites—the aluminum industry (perhaps coincidentally) has been the focus of attention. Past discussion of this matter mainly concerned the bauxite mines at Weipa and Gove, where the claims of sacred sites have been intertwined with claims by the aboriginals for royalties from the mined bauxite. A clearer case has emerged at Alcoa's proposed Portland smelter, as Alcoa's Environmental Effects Statement, 1980 indicates:

> Most of the recorded archaeological sites were outside of the Alcoa smelter site boundary. Construction of the proposal would probably destroy or cover . . . recorded sites. . . . However, none of the sites affected is significant. It is possible that more archaeological sites could be uncovered during construction. Alcoa will engage an archaeologist to determine and document any sub-surface sites in advance of any major earth work activity.[25]

The state of Victoria, in which Portland is located, provides protection for significant relics and sites through the Archaeological and Aboriginal Relics Preservation Act of 1972. There has been considerable difficulty, however, in designating sites of "significance" and establishing who should have legal standing in the matter to bring suits. Two aboriginal women took action through Victoria's Supreme Court under this legislation in a bid to prevent work on the smelter, but they were ruled not to have legal standing in the matter and the case was dismissed. More recently, the Gunditj-Mara descendants living in the Portland area have had discussions with Alcoa to identify areas to which they attach special significance, and it seems there will not be a major conflict with Alcoa's development plans.

Yet many ambiguities exist in the state laws protecting aboriginal archaeological sites, and the federal government is currently proposing new uniform legislation to cover all states and to make a clearer statement of the procedures to be followed in resolving such issues. Present law is such that, in the case of major development projects, it is possible that descendants of the relevant tribe, although dispersed and no longer occupying the site in question, can take court action to forestall or hold up a project such as the Portland smelter. This circumstance may have given rise to the impression in the public mind that political claims of a wider kind are pushed through under the guise or mechanism of protecting a sacred site. Although so far the courts have not agreed with these claims, it could be that in the next case, or under new legislation, the courts may reverse their previous positions and side instead with the claimants.

CONCLUSIONS

The year 1980 was one of great promise for the Australian aluminum industry as the recognition dawned that the international energy crisis offered Australia the chance for a major comparative advantage in aluminum smelting. Here was a country with vast reserves of both coal and bauxite. It remained only to construct power stations and smelters, and the industry would be internationally competitive. The great excitement of 1980 was short lived, however, and by 1984 it had all but evaporated. Considerable construction of new capacity did occur, but it has fallen far short of the many forecasts for the industry made in the late 1970s. The answers to two questions are central to understanding what has happened: (1) Did the two energy crises really make Australia internationally competitive for the location of new aluminum smelters? and (2) To what extent was international competitiveness squandered by ineffective economic policy?

In terms of the first question, it is clear that the sharp increases in oil prices in 1973 and 1979 made coal an attractive fuel for electricity generation and caused Australian electricity to be relatively less expensive than it had been in the past. But it does not automatically follow that Australia should have gone out to seek energy-intensive industries. An examination of the issues involved in the pricing of coal for electricity generation make this clear. Because steaming coal can be exported as well as converted into domestic electricity, a computation of competitiveness that reflects true opportunity costs must assign an export parity price to the coal used for domestic electricity production. Because government electricity authorities tend to price coal at its extraction cost in their cost allocation procedures, however, electricity tariffs are set below true opportunity costs. It remains doubtful whether issues such as these have been considered by public officials responsible for electricity tariff setting. Concepts such as opportunity costs and export parity pricing are likely to be still seen as the artificial constructs of academic economists and thus not pertinent to real decision making. One must have some sympathy with this view, especially given the considerable conceptual and practical difficulties of establishing a fair figure for the export parity price of steaming coal. Nevertheless, the essential idea remains that Australia should not be regarded as having a competitive advantage for new smelter construction if those smelters are not economically viable when charged electricity prices that reflect the price that could be obtained for coal sold on world markets.

It is possible that Australia did not at any time have a significant competitive advantage in aluminum smelting. It is also possible that today electricity tariffs are artificially low as a result of gross errors in the cost allocation procedures that led electricity authorities to set tariffs that do not reflect the true cost of capital being employed in electricity generation. As was noted earlier in this chapter, in at least one case a state electricity authority had established a required real rate of return on capital of minus 3 percent. In addition, other practices, such as allocating the interest cost of capital in construction to other unrelated operating accounts, result in a significant understatement of the true capital stock on which some required rate of return must be earned. The commercial secrecy surrounding the activities of state electricity authorities in Australia makes it extremely difficult to formulate a reliable assessment of the extent of these problems. There is a clear need for legal reforms that would significantly increase the disclosure requirements of state authorities. Only careful public scrutiny can ensure that the resources of the taxpayer are not being squandered by bureaucrats pursuing policies that do not consider the scarcity of the resources to be exploited.

It is possible that Australia could have achieved a competitive advantage in aluminum smelting in 1980, but by 1983 any hopes for such an advantage had been significantly reduced. The best explanation for such a course of events is that excessively high wage settlements and their associated industrial disputes made many new projects uneconomic. A shift in competitive advantage always creates economic rents and a scramble among market participants to capture those rents. It is certain that in Australia labor was most successful in locking in the economic rents of the resource boom, but it is equally possible that the demands exercised through the monopoly power of the labor unions extracted more than a fair share of the potential new economic rents and in fact killed the boom.

The high labor costs on new power stations and smelter projects probably made them uneconomic, thus victimizing the labor force and the economy at large when the projects could not proceed. Chastened by the experience, a new era of conciliation has settled on the Australian trade unions. In 1984 industrial disputes were at their lowest level since 1969, and wage increases have been significantly moderated. It is now necessary to make some fundamental changes in the procedures for settling industrial disputes. There is a need to rationalize unions on an "industrial" rather than a "trade" basis to reduce the incidence of demarcation problems. And there is also a need either to limit the use of collective bargaining, or to have it closely regulated, to stem the practice of wage-gouging that so often has occurred on major construction projects once they are partially completed.

Finally, environmental safeguards and the protection of aboriginal sacred sites are issues that must always be considered in broaching new development projects. Neither should present major obstacles, however, to the future reinvigoration of the Australian aluminum smelting industry. In addition, new, uniform aboriginal land rights legislation seems likely to remove the remaining uncertainties in these areas.

In the end, was its competitive advantage squandered, or was there never a true comparative advantage in aluminum smelting for Australia? The case has not yet been resolved. Perhaps the final answer borrows something of both possibilities: the true competitive advantage was not as large as it first seemed, and what competitive advantage existed was squandered.

NOTES

1. For further discussion, see Department of Industry and Commerce, "The Development of the Bauxite, Alumina, and Aluminum Industries," submission to the Senate Standing Committee on Natural Resources (Canberra, 1981); T. K. McDonald, "The Development of the Aluminum Industry in the Pacific Region," presented at the 15th Pacific Science Conference, Wellington, New Zealand, 1983; M. G. Porter, "The Aluminum Industry in Australia in the 1980s" (Centre of Policy Studies, Monash University, 1981); and M. R. Rayner, "The Economics of the Australian Aluminum Industry," presented at the Australian Institute of Mining and Metallurgy Conference, Melbourne, 1982.

2. Arm's-length transactions are those in which the buyer and the seller are not owned or controlled by a common third party.
3. For further discussion, see the Senate Standing Committee on Natural Resources, *Report on the Development of the Bauxite, Alumina, and Aluminum Industries* (Canberra, Australian Government Publishing Service, 1981).
4. For further discussion of electricity pricing in relation to the aluminum industry, see R. H. Burke and G. H. Cranby, "An Analysis of the Impact of the Aluminum Industries on Australian Energy Supply–Demand Balance," Comalco Energy Research Paper No. 2 (1981); and Department of Treasury, "Energy Markets—Some Principles of Pricing," Treasury Economic Paper No. 5 (1979).
5. See Department of Treasury, *The Development of Bauxite, Alumina, and Aluminum Industries.* Submission to the Senate Standing Committee on Natural Resources (Canberra, Commonwealth Government Printer, Official Hansard, 1981) pp. 1007–1008.
6. Ibid., pp. 64–68.
7. H. W. Dick, "Power Subsidies to Aluminum Smelters in NSW." Department of Economics Discussion Paper No. 18 (University of Newcastle, 1981).
8. Ibid., p. 8.
9. For a discussion of the supply of electricity to this smelter, see the contract consultants' report by D. R. Gallagher, G. D. McColl, M. C. Copeland, and T. K. McDonald, "The National Economic Benefits of the Gladstone Aluminum Smelter." (Melbourne, Comalco Limited).
10. "Electricity Concern," *Australian Financial Review,* August 4, 1981, p. 7.
11. P. L. Swan, "Pricing of Electricity to Alcoa at Portland, Victoria." Australian National University Working Papers in Economics and Econometrics No. 54 (Canberra, 1981) p. 5.
12. See Department of Treasury, *The Development of Bauxite,* p. 586.
13. T. E. Copeland and J. F. Weston, *Financial Policy and Corporate Policy* (Reading, Mass., Addison-Wesley, 1979).
14. P. L. Swan, "Pricing of Electricity," p. 8.
15. State Electricity Commission of Victoria, "Loy Yang Project: Report on Proposed Extensions to the State Generating System" (Victoria State Government, 1967) p. 38.
16. For further discussion, see Swan, "Pricing of Electricity"; and Porter, "The Aluminum Industry in Australia."
17. By way of contrast, investments in bauxite mines of more than $5 million must have a minimum of 50 percent Australian equity with at least 50 percent of the voting strength of the board or controlling body held by Australian interests.

18. Recently, sections 136 and 260 of the Income Tax Assessment have been revised following a court decision in *Federal Commissioner of Taxation* v. *Commonwealth Aluminum Corporation, Ltd.* (CAC) (80 ATC4371).
19. Export controls on bauxite and alumina were introduced in 1973 as part of the Australian customs regulations ("Prohibited Exports"). The extent of government interference in export contracts through these regulations has been inconsequential, however.
20. R. C. Gregory, "Some Implications of the Growth of the Mining Sector," *Australian Journal of Agricultural Economics* vol. 20, no. 1 (1976) pp. 1–42.
21. There are no penal clauses in Australian industrial law that pertain to union leadership.
22. In 1974 the Australian government unexpectedly imposed a tax of $10 per ton on exported coal.
23. The most spectacular wage gains were in coal mining, a phenomenon that has a long-run bearing on the price of electricity to smelters.
24. The labor supply issues are covered extensively in the report of the Department of Labour Advisory Committee, *Prospective Demand for and Supply of Skilled Labor, with Particular Reference to Major Development Projects*. Report of the Working Party (Canberra, Australian Government Publishing Service, 1981).
25. Alcoa of Australia, "Alcoa Portland Aluminum Smelter, Environmental Effects Statement" (Melbourne, 1980) p. 93.

6

BRAZIL: THE TRANSITION TO AN EXPORT INDUSTRY

ELIEZER BRAZ-PEREIRA

The Brazilian aluminum industry[1] is an example of a fledgling industry that has matured and become a major world player. Until the early 1970s, aluminum smelting occurred for the most part in the industrialized countries examined in this volume—the United States, Japan, and Western Europe—each of which was a major consumer of aluminum. (Canada, the sole exception, was a major producer but not a major consumer of aluminum.) The less developed countries or LDCs—a category that includes Brazil—had smelters that largely served local markets. The production of aluminum for export in LDCs was only found in those cases in which very favorable resource availability allowed the LDCs to overcome their location disadvantages vis-à-vis the industrialized countries: the distance from consumers, the scarcity of skilled labor, the lack of infrastructure, and the perception of higher risks for any investment. Thus, the only two exporting LDCs were Surinam, with both bauxite and hydropower, and Ghana, with hydropower. After the first oil shock, however, export-oriented aluminum smelting developed in several LDCs, signaling a change from the previous pattern of locating smelters close to the consumers in industrialized countries. The main reason for this change was the increasing scarcity of low-cost electricity in the industrialized countries, a situation that has continued to this day.

Eliezer Braz-Pereira is in the Department of Minerals and Geology at the Federal University of Paraiba, Brazil.

Among the LDCs, Brazil seems to have the potential for the greatest expansion of smelting capacity. When its current projects are completed, Brazilian smelting capacity is expected to increase from 377,000 tons in 1982 to 734,000 tons in 1987. Brazil is unique, at least among LDCs, in combining a large, unutilized hydro potential with ample bauxite ores, a well-developed industrialized base, and a government policy favorable toward aluminum production. Brazil's industry thus is a good case study of the process by which nontraditional producers become internationally competitive.

The analysis of the Brazilian case is complicated, however, by the existence of two distinct sectors in the aluminum industry. The older southern sector serves local markets and is an example of the aluminum industries found in some LDCs prior to the 1970s. The newer northern sector is an example of the export-oriented industry that is now emerging in some LDCs. Figure 6-1 shows the location of smelters in Brazil and illustrates the geographic division of the two sectors.

PRIMARY ALUMINUM PRODUCTION IN SOUTHERN BRAZIL

A Historical Perspective

The first attempt to smelt aluminum in Brazil was made in 1945 by Eletroquímica Brasileira S/A (Elquisa), which established a single small smelter with an annual capacity of 2,000 tons. The company closed shortly after it opened, however, because it was unable to compete with imports of low-cost aluminum that were available from the wartime expansion of capacity in the United States. In 1951, in a move that marked the start of the domestically oriented industry, Alcan bought Elquisa and renamed it Alcan Alumínio do Brasil. From the beginning, Alcan's operation was fully integrated with the mining, refining, and smelting processes all located at Ouro Preto in the state of Minas Gerais. The initial production was small, but it increased over the years as capacity expanded (table 6-1).

The next entrant to the industry was Companhia Brasileira de Alumínio (CBA) in 1955. This company, controlled by the Brazilian Votorantim group, started production at Mairinque in the state of São Paulo where it also had an alumina refinery and fabricating facilities. CBA obtained its bauxite from mines at Poços de Caldas in

Figure 6-1. Location of the aluminum industry in Brazil. *Source:* Based on maps presented by Carlos Romano Ramos, "Perfil Analitíco do Alumínio" (Aluminum Analytic Profile), Departamento Nacional da Produção Mineral, Boletim 55 (Brasília, Editora Grafisa, 1982) pp. 140 and 145. Additional information included by the author.

the state of Minas Gerais and its electricity from its own hydro generating plant at Sorocaba in the state of São Paulo. As table 6-1 shows, CBA had achieved the same output level as Alcan by 1959, and after that year the two companies had equal market shares. The combined production of Alcan and CBA grew nearly sixteenfold from 2,699 tons in 1955 to 43,294 tons in 1969.

Alcoa became the third producer of primary aluminum in Brazil by establishing a subsidiary in 1970, Alcoa Alumínio. The subsidiary was a joint venture with Hanna, a U.S. mining company;

Table 6-1. Domestic Production of Primary Aluminum in Southeast Brazil and Imports (metric tons)

Year	Alcan	Alcoa	CBA[a]	Valesul	Total domestic production	Imports[b]
1951	400	—	—	—	400	15,544
1952	1,100	—	—	—	1,100	11,007
1953	1,200	—	—	—	1,200	11,805
1954	1,500	—	—	—	1,500	17,495
1955	1,700	—	999	—	2,699	6,705
1956	1,700	—	3,805	—	5,505	14,194
1957	2,100	—	4,686	—	6,786	13,260
1958	2,700	—	6,654	—	9,354	14,307
1959	6,500	—	7,722	—	14,222	9,312
1960	7,400	—	7,573	—	14,973	15,015
1961	9,600	—	8,270	—	17,870	18,476
1962	13,000	—	7,979	—	20,979	19,507
1963	13,500	—	6,558	—	20,058	25,815
1964	14,600	—	11,439	—	26,039	18,549
1965	15,400	—	14,163	—	29,563	21,844
1966	17,200	—	15,734	—	32,934	39,540
1967	19,300	—	18,775	—	38,075	28,014
1968	22,123	—	19,301	—	41,424	33,601
1969	22,824	—	20,100	—	42,924	44,795
1970	25,129	7,900	23,118	—	56,147	27,433
1971	27,205	24,942	28,500	—	80,647	23,060
1972	35,853	31,281	30,502	—	97,636	44,028
1973	41,717	30,407	39,616	—	111,740	51,700
1974	45,534	29,513	38,551	—	113,598	104,788
1975	55,600	29,967	35,778	—	121,345	78,804
1976	59,335	41,296	38,543	—	139,174	78,101
1977	59,283	59,458	48,368	—	167,109	82,756
1978	61,355	58,971	66,039	—	186,365	60,435
1979	79,684	82,304	76,144	—	238,132	51,816
1980	87,930	89,311	83,370	—	260,611	46,702
1981	87,345	88,537	80,536	—	256,418	28,241
1982	88,526	89,674	96,636	24,218	299,054	10,766

Source: Associação Brasileira do Alumínio (ABAL), *Anuário Estatístico ABAL 1982* (ABAL Statistics Yearbook) (São Paulo, ABAL, 1983) p. 8.

[a] CBA = Companhia Brasileira de Alumínio.

[b] Primary metal and alloys.

Hanna took a 32 percent interest, and Alcoa took the rest. Unlike the first two entrants, however, Alcoa started up at a much more efficient scale (30,000 tons per year) and became the primary supplier to Brazilian independent fabricators rather than integrating forward into fabricating.[2] These independent fabricators also relied on imports, which they bought at above the international price because of tariffs. As a result, fabricating prices were higher in Brazil, which made the fabricating operations of Alcan and CBA more profitable.

Almost from the outset, then, Alcoa's production was at about the same level as that of the other two producers, and they grew at comparable rates, dividing the market about equally among them. It has been suggested that some sort of market-sharing agreement might have existed among the three producers.[3] It has also been said that these three companies failed to expand production enough to eliminate imports and specialized among themselves in serving different types of customers.

In 1972, Alcan, through its subsidiary Alumínio do Brasil Nordeste (Alunordeste), built its second smelter in Aratu in the state of Bahia in northeastern Brazil. It was the first smelter outside southeastern Brazil and the first to rely on imported alumina. It had the advantages of a coastal location with easy access to imported alumina and consumers and, most importantly, of being eligible for the government incentives offered to firms locating plants in northeastern Brazil. Those incentives allowed Alcan to offset 50 percent of the income tax due on its activities elsewhere in Brazil against its investment in Alunordeste.

All of these smelters were aimed exclusively at supplying the growing domestic market. Growth in aluminum consumption in the 1970s averaged 12.7 percent annually, raising consumption from 96,583 tons in 1970 to 367,492 tons in 1979 (table 6-2).

Valesul, the fourth entrant in 1982, was established as a completely nonintegrated facility. The smelter was built near Rio de Janeiro despite the cost advantages at the time of locating in the north. It relied on imports of alumina, and it sold primary aluminum to independent fabricators. In the future, it will receive alumina from Alunorte, an alumina plant under construction in the Amazon region. With Valesul in operation, Brazil for the first time could supply all its domestic demand and still have an exportable surplus. Even though depressed economic conditions had much to do with the creation of the surplus, its occurrence nevertheless marked the beginning of Brazil's presence in the international market as a supplier of primary aluminum. Valesul may not continue as

Table 6-2. Consumption of Aluminum in Brazil, 1970–1979 (metric tons)

Year	Consumption
1970	96,583
1971	129,093
1972	162,203
1973	200,357
1974	255,733
1975	252,874
1976	262,144
1977	322,820
1978	312,168
1979	367,492

Source: Departamento Nacional de Produção Mineral (DNPM), *Balanço Mineral Brasileiro* (Brazilian Mineral Balance) (Brasília, DNPM, 1980) p. 48.

an aluminum exporter, however, if the growth of domestic demand resumes its earlier rate. Should this occur, the southern industry will return to serving exclusively domestic markets.

Comparative Advantage

In the 1950s, when Alcan and CBA started their production of primary aluminum in southeastern Brazil, the region offered favorable conditions for aluminum smelting. Three important inputs were locally available: alumina, electricity, and labor. Alumina could be produced from the bauxite deposits of the state of Minas Gerais; the region's hydroelectric potential was very large, offering good possibilities for low-cost electricity; and the region already had considerable industrial development and an adequate supply of skilled labor and managerial personnel. Because aluminum consumption was concentrated in the Southeast, local production of aluminum minimized transportation costs. Domestic production had, of course, a transportation cost advantage over imports.

Furthermore, both capital and technology were available on favorable terms because Alcoa and Alcan, as owners of two of the producers and members of the big six, had access to both capital and technology. CBA was owned by Votorantim, the largest private Brazilian group, and the availability of capital does not seem to have limited its growth. CBA purchased technical assistance first from Elken, then from Montecatini, and finally from Pechiney.[4] Valesul, the fourth entrant, was financed by the government and by Shell, the international oil company. Technology was supplied by Reynolds, another of the big six.

The major disadvantage of the southern industry was its small market, as a result of which production was carried on at less than the minimum efficient scale. As table 6-1 indicates, in 1951 when Alcan started annual production, apparent consumption (domestic production plus imports) was only 15,944 tons. In 1956, one year after CBA entered the industry, apparent consumption was still only 19,699 tons, whereas the minimum efficient size for a smelter was 100,000 tons. The small size of the market thus prevented the realization of economies of scale. In fact, production at any level close to an efficient scale in the Southeast would have generated a surplus. Exports of that surplus would have faced high transportation costs from the Southeast to the main export markets in the United States, Japan, and Western Europe.

Consequently, even though southeastern Brazil had a favorable combination of resources for aluminum smelting, the small size of the market created a cost disadvantage for the industry. Table 6-1 shows that in the early 1970s all producers had increased their annual output to around 30,000 tons and that in the early 1980s each of them was producing close to 100,000 tons per year (Alcan had two smelters). This level of production means that the industry in the Southeast was in a position to realize economies of scale, thus overcoming its main disadvantage.

By about 1973, Brazil's aluminum industry was internationally competitive based on the cost schedule shown in table 6-3. (The total costs for 1973 were $530 per ton for a smelter in the Southeast. With the international market price at that time at $770 per ton and the posted list price at $660 per ton, Brazilian costs were less than the world price by a significant margin.) The relationship of costs and revenue is confirmed by returns on equity of 15 and 23 percent for 1979 and 1980, respectively, for the southern smelters, even though Brazilian domestic prices for primary aluminum were set by the government at less than world prices.[5] Another indication of Brazil's international competitiveness, which has continued up to the present, is that fabricators in the southeastern region have been able to export their products, which are made from locally produced primary aluminum. Yet table 6-3 also shows that in 1973 the southern industry was at a cost disadvantage compared to proposed smelters in the North.

Table 6-3. Costs of Production of Primary Aluminum, 1973 (dollars per metric ton)

Location	Southeast[a]	Northeast[b]	North[b]
Electricity	132.00[c]	132.00[d]	77.50[e]
Alumina (1.93 tons at $90/ton)	193.00	193.00	193.00
Capital costs	56.00	40.00	48.00
Petroleum coke (0.4 ton at $54/ton)	21.60	21.60	21.60
Aluminum fluoride	11.30	11.30	11.30
Cryolite	9.20	9.20	9.20
Pitch	3.10	6.20	6.20
Carbon blocks	8.40	8.40	8.40
Other expenses	10.00	20.00	20.00
Labor	50.00[f]	37.50[g]	50.00[h]
Maintenance	36.00	18.00	18.00
Total	530.00	489.20	463.20

Source: Raymundo C. Machado, *Alumínio Primário no Brasil* (Primary Aluminum in Brazil) (Ouro Preto, Minas Gerais, Fundação Gorceix, 1982) p. 163.

[a] 30,000 tons/year.
[b] 100,000 tons/year.
[c] 16,000 kWh at 8 mills/kWh.
[d] 15,500 kWh at 8 mills/kWh.
[e] 15,500 kWh at 5 mills/kWh.
[f] 50 worker-hours at $1.00/hour.
[g] 25 worker-hours at $1.50/hour.
[h] 25 worker-hours at $2.00/hour.

Even if the existing smelters in the Southeast are now internationally competitive, it seems likely that the region was never a low-cost location for aluminum smelting. In the 1950s and 1960s domestic demand was too small for efficient production; in the 1970s growth in demand eliminated that problem, but the same process of economic development that increased the demand for aluminum also increased the demand for electricity. Low-cost sources of electricity were fully committed, and the marginal social cost of electricity began to rise sharply.

According to the Ministério das Minas e Energia (MME), during 1969–1979 electricity consumption grew at an average rate of 13.4 percent.[6] This high rate of growth reflects rising incomes, an increase in population, and greater urbanization. (In the decade of the 1970s, gross domestic product increased at an annual rate of 8.9 percent, and population increased at an annual rate of 2.9 percent. The share of urban population, already 56 percent in 1970, also continued to rise.) Such rapid growth in electricity consumption is expected to continue. MME estimates that electricity consumption will increase at an annual rate of 11.4 percent in 1980–1985, given the still very low per capita consumption and Brazil's continued efforts to extend electric power to the 42 million people not yet served.

Electricity in Brazil historically has been generated from hydropower. In the 1970s that source accounted for about 92 percent of electricity generation.[7] Installed electricity generating capacity in December 1979 was 24.1 gigawatts; the goal for 1985 was 50 gigawatts, which is still only half of the estimated uninterruptible hydropower potential of 106.5 gigawatts (based on an average low-rainfall year). Brazil's overall estimated hydropower potential is expected to meet the nation's electricity demand until the year 2000.

Table 6-4 shows the consumption of electricity and the hydropower potential for the different regions of the country. Almost half of the hydropower resources are found in the North, and these are almost untouched. On the other hand, in the Southeast and Northeast, where most electricity consumption takes place, a high portion of the available hydropower was already in use by 1980. Use of the remaining hydropower in these two regions, however, entails higher electricity costs for the power generated from these sites because the sites are farther from consuming centers and have unfavorable construction conditions. In contrast to the North, competition among consumers in the Southeast and Northeast increases the marginal social opportunity costs of the electricity used in those regions.

By 1984 there were optimistic indications of electricity demand in the Southeast, leading to the development of what is now surplus capacity. That capacity includes two recently completed projects: the first nuclear plant, Angra I, in the state of Rio de Janeiro, and the Itaipu hydro plant, a joint venture between Brazil and Paraguay. Because the power surplus is expected to be temporary, however, the unavailability of low-cost power over the long term means that neither new aluminum smelting capacity nor major expansions of existing capacity are to be expected in the Southeast.

Public Policy

The protection provided by public policy was a decisive factor for the development of aluminum smelting in southeastern Brazil. Infant import-substitution industries in LDCs usually have costs that

Table 6-4. Electricity Consumption and Hydropower Potential in Brazil, 1980

Region	Consumption		Hydropower potential[a]		Hydropower potential[b]	
	Gigawatt-hours	Percentage	Installed or under construction	Unused	Installed or under construction	Unused
North	2,293	2	4.3	95.7	2.0	44.4
Northeast	15,788	13	45.2	54.8	3.1	3.7
Southeast	84,024	70	44.8	55.2	11.2	13.8
South	15,006	12	29.3	70.7	6.4	15.4
Center–South[c]	3,615	3	—	—	—	—
Total	120,726	100	—	—	22.7	77.3

Source: Ministério das Minas e Energia, *Modelo Energético Brasileiro* (Brazilian Energy Model) (Rio de Janeiro, Gráfica e Editora Celsus Ltda., 1981) p. 56.

[a] Relative to the region (percentage).

[b] Relative to the country (percentage).

[c] The hydropower of the Center–South is included in the North and in the Southeast.

are higher than the price of imports and often receive protection from the government in order to develop in the face of lower priced imports. Protection is most commonly accomplished by a tariff on imports. Figure 6-2 shows the situation facing the southern aluminum industry in 1951. The product of the high-cost infant industry (S_d) could not compete with imports (S_m) in the free trade situation. With the imposition of a tariff T on imports, supply shifts up to $S_m{}^1$ and domestic production becomes feasible. Demand is met by domestic production (OQ_1) and by imports (Q_1Q_2).

As domestic demand for aluminum grew, output could increase to realize economies of scale, thus improving the international competitiveness of the industry. (The realization of economies of scale alone would not justify protection. If producers take a long view and can borrow freely, the firms themselves can recover their initial higher costs from the profits when the firm reaches an optimal size.[8]) Although the protection policy was successful in allowing the development of aluminum smelting in the Southeast, it is not clear whether the social benefits generated by the industry justified the costs of such protection.

These costs are difficult to estimate because the protective measures adopted by the government have been complex and have been changed often. An ad valorem tariff of 37 percent has been in effect

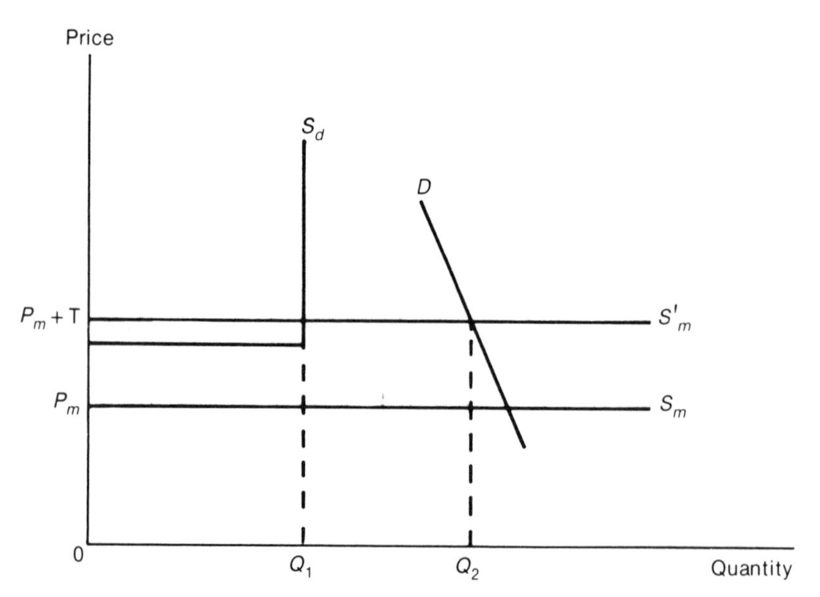

Figure 6-2. Protection of an infant industry by a tariff.

since the outset of the industry. Yet until 1982—when for the first time Brazil had an exportable surplus of primary aluminum—the tariff was always waived. Instead, other measures were used to limit imports: the linkage of import licenses to purchases of domestic metal, the concession of exclusive import rights,[9] the creation of a company to import and market aluminum,[10] a 25 percent tax on financial transactions involving imports, and a compulsory deposit equal to the value of the imports (in effect, a twelve-month loan to the government without interest).

Figure 6-3 shows a situation in which the domestic industry has become competitive with imports without protection. Domestic supply is represented by S_d and imports by S_m. Domestic producers would supply OQ_1, and imports would add up to Q_1Q_2. As already discussed, there are indications that these were the circumstances of aluminum smelting in the Southeast during the 1970s. With the growth in demand, scale economies were realized, and domestic producers then had a cost advantage over imports in terms of lower transportation costs because of the local availability of bauxite resources and proximity to consumers. Yet protection has continued despite the new competitive position of domestic producers.

Until 1982 the expansion of domestic production always lagged behind demand with imports filling the gap. Low-cost domestic production allowed domestic producers to collect the economic rents generated by limiting their production to a proportion of the market. Expanding capacity and production to eliminate imports might have lowered prices to consumers sufficiently to reduce their profits. In Figure 6-3, for example, with a domestic supply of S_d that is equal to all the domestic demand, price would decrease from P_m to P_d. Keeping production at a volume that was less than demand also meant that domestic producers would be insulated from fluctuations in demand because all the adjustment to such changes would occur in the volume of imports. Hence, smelters could operate at their optimum rates, lowering their costs.

Protection was only one public policy measure; two others of significance were policies for the expansion of hydropower and price controls. The development of electricity generation from abundant hydro sources allowed the industry to grow without the limitations imposed by a scarcity of electricity. Aluminum producers in the Southeast and the Northeast, however, did not have special electricity rates but instead paid the same rate as other large industrial consumers. They did benefit, however, from a 98 percent reduction in the compulsory loan that electricity consumers had to make to

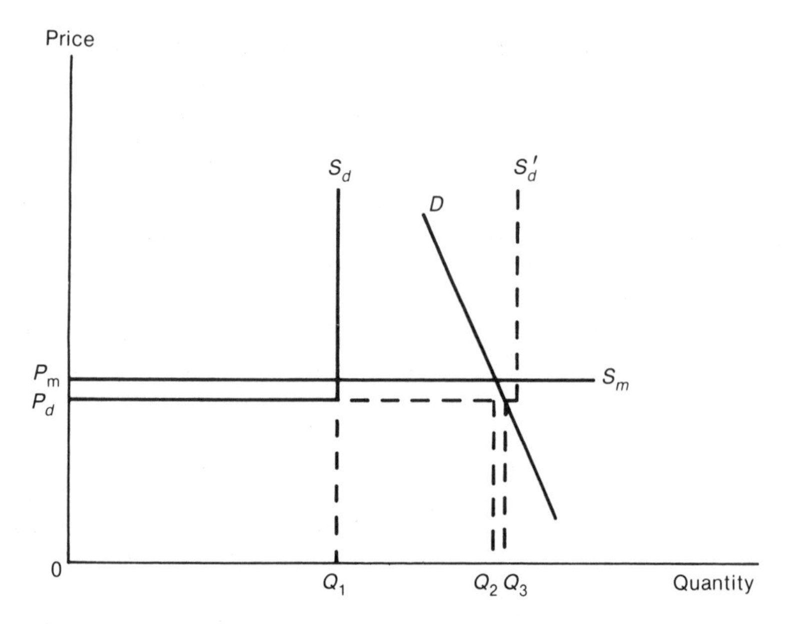

Figure 6-3. The mature aluminum industry.

the government during 1962–1982.[11] (The loan was defined as a percentage of the monthly electricity bill and was added to it.) Although this benefit was small, other large industrial consumers were not given this concession.

During the 1970s industrial growth increased competition for the electricity used by the aluminum industry. In addition, the energy crisis brought about energy substitution programs in sectors such as transportation, further increasing the demand for electricity. As a result, electricity prices to aluminum smelters increased from 6 to 14 mills.

Since 1964 the domestic price of primary aluminum has been controlled by the Brazilian government through the Conselho Interministerial de Preços (CIP). Figure 6-4 shows CIP prices during 1970–1983 compared to U.S. spot prices and Alcan list prices (which tend to be the international prices, given Alcan's major role as an exporter). For most of the period, domestic rates followed the trend of the two international prices but were somewhat higher. In 1973–1974 domestic prices were lower than spot prices because the surge in prices that occurred elsewhere in those years was not matched in the Brazilian domestic market. Since 1979 the behavior of Brazil's domestic prices has been quite different from those of the

Figure 6-4. Aluminum prices during 1970–1983. *Source:* R. C. Machado, written communication.

Alcan list price and the spot price. The large price declines in 1980 and 1983 were the result of the two 30 percent devaluations of the cruzeiro that took place in December 1979 and February 1983. After the decline of 1980, prices reacted, once again following the trend of the Alcan list price. In 1983 there was an increase in domestic prices after adjustment to the devaluation.

From February 1982 to February 1983 price controls were eliminated, but the period is too short for an appraisal of the change in policy. Prices initially rose and then declined, perhaps as a reaction to the entry of Valesul into a very depressed market.

Domestic prices have been based on cost data submitted by the industry to CIP. The fact that those prices for most of the 1970–1983 period were higher than world prices might be taken as an indication that domestic costs were high and therefore that the industry was not internationally competitive. The price differences were usually minor, however, and other evidence indicates that the aluminum industry in the Southeast was internationally competitive by the 1970s.

THE ALUMINUM INDUSTRY IN NORTHERN BRAZIL

Whereas the expansion of aluminum smelting in the Southeast was driven by a growing domestic demand, the development of the industry in the North was driven by the abundance of natural resources. A further distinction is that the northern industry at its outset was directed toward producing aluminum for export.

The Development of the Industry

After 1973 the potential for the development of aluminum smelting for export in the North was widely recognized. The oil price rise during that year sharply affected the competitive position of many smelters worldwide and especially in Japan. The Japanese response, which was described in chapter 4, was to shut down inefficient smelters at home and participate in the development of new capacity abroad. This strategy was reflected in Japanese participation in such projects as Albrás in Brazil, Asashan in Indonesia, and Venalum in Venezuela.

Even before these events, however, Brazil had begun to pursue its development of the industry. In 1972 very large deposits of bauxite were discovered near the Trombetas River, a tributary of the Amazon. With this discovery, it became clear that Brazilian reserves of bauxite were second only to those of Guinea and Australia. Government interest in exploiting the Trombetas deposits led to the formation of Mineração Rio do Norte (MRN)—a joint venture of the Companhia Vale do Rio Doce (CVRD) (46 percent), Alcan (19 percent, later expanded to 24 percent), CBA (10 percent), and other foreign companies (Shell, Norsk Hydro, and Reynolds). MRN was formed to mine bauxite near the Trombetas River. The involvement of the government-controlled CVRD, a large iron ore producer, represented the diversification of that company into the aluminum industry.

All bauxite to be produced in the first phase of the Trombetas project, which began producing in 1979, was exported to the foreign partners of MRN. Expansion plans were developed, however, to provide for some of the bauxite to be supplied to alumina refineries that were to be constructed in the Amazon region.

The involvement of CVRD in bauxite mining led the company to consider integrating into both alumina refining and aluminum smelting. As an exporter of iron ore, CVRD already had a long-standing relationship with the Japanese, who were interested in new sources of aluminum. These mutual interests and relationships led to CVRD's partnership with a group of Japanese aluminum producers in the Albrás project. The project, which was planned to produce 600,000 tons of aluminum per year, was to be located near Belém, at the mouth of the Amazon River. The supply of electricity was to be guaranteed by the decision of the government to build the Tucuruí plant on the Tocantins River, another tributary of the Amazon.

Negotiations between CVRD and the Japanese in 1976 resulted in a reduction of the smelter's planned output and the separation of the initial project into two ventures, Albrás and Alunorte. Albrás became a joint venture of CVRD (51 percent) and Nippon Amazon Aluminum Company (NAAC) (49 percent). (NAAC is a consortium that includes the Japanese Overseas Economic Cooperation Fund or OECF with a 40 percent share, and 32 private firms, including producers of primary aluminum, fabricators, consumers, trading companies, and a bank.[12]) Alunorte is also a joint venture of CVRD (61 percent) and NAAC (39 percent). Both projects were to begin operations in 1980.

Subsequently, however, the start-up of Albrás was postponed until 1985 and that of Alunorte until 1990. The two projects were delayed by difficulties in obtaining financing from Brazilian sources, the unavailability of electricity before 1984, and controversy in the negotiations between the Japanese and Brazilian partners that extended to both projects. The opposition that developed from other groups in Brazil was based on several factors: (1) the Japanese were minority shareholders but had veto power on all technical decisions concerning the project; (2) NAAC appointed the technical, planning, and controlling directors; (3) the project would receive electricity at favorable rates and sell aluminum to Japanese consumers at 6 percent below the Alcan list price; (4) profits accruing to NAAC would not be taxed in Brazil; and (5) all relevant matters required a two-thirds quorum for approval.[13] Still, construction has actually begun on both projects. NAAC will take all the aluminum produced in the first phase of operations (160,000 tons) and 50 percent of the production in the second phase.[14]

There have also been other projects developed—for example, Alumar, which involves Alcoa Alumínio (60 percent) and Billiton Metais (40 percent), a subsidiary of Shell. Unlike the two ventures discussed above, Alumar in its first phase is planned to produce 500,000 tons of alumina and 100,000 tons of aluminum annually. The plants are located on the island of São Luís, close to the capital of the state of Maranhão. Electricity comes from the Tucuruí plant, and bauxite will be supplied by MRN until Alcoa develops its large bauxite concessions in the Amazon region. These mines are expected to be in operation by the end of the 1980s. Alumar will supply alumina to Albrás until Alunorte starts operations. In 1984 the Brazilian Camargo Correa group acquired an interest in Alumar, and as a result the smelter is scheduled to expand its capacity from 100,000 to 200,000 tons annually. Eventually, Alumar will reach a capacity of 350,000 tons per year.

Alumar is part of Alcoa's strategy for developing new smelting capacity in cost-competitive locations that can supply a growing domestic market (such as Brazil) and still have a surplus for export to meet some of Alcoa's incremental needs in the United States. As for Shell, with Alumar the company significantly increased its participation in the aluminum industry in Brazil.

International Competitiveness

Electricity is such an important component in the production cost of primary aluminum that after the oil price shock, locations with abundant hydroelectric generating potential, such as the Amazon region, acquired a major competitive advantage. Abundant hydropower, however, is a necessary but not sufficient condition for an internationally competitive aluminum industry. For example, several LDCs that appear to have cheap hydropower nevertheless have not developed a competitive industry. Other factors, then, become important in defining where aluminum smelting will be located; of particular importance in Brazil was the abundance of bauxite and a favorable government policy.

Late in the 1960s the discovery of extremely large deposits of bauxite in the Amazon region changed the focus of bauxite mining from the Southeast to the North. Table 6-5 shows the distribution of bauxite reserves in Brazil. Total reserves are more than 5 billion tons, with 98 percent in the state of Pará and only 1.6 percent in the state of Minas Gerais, where bauxite mining has been traditionally located.

The bauxite deposits of the Amazon region also changed the ownership pattern of the mining industry. Bauxite deposits in the Southeast are controlled by the three fully integrated companies—that is, Alcan, Alcoa, and CBA. The discovery of the deposits of the Amazon opened up possibilities for other companies to acquire important bauxite holdings. CVRD in particular seized the opportunity to become the largest owner of bauxite reserves while Alcoa expanded its holdings to become the second largest. Other active companies include CBA and Shell Brasil. The shares of Alcan, Reynolds, and Norsk Hydro stem from their participation in MRN (table 6-6).

An abundance of bauxite, however, is not a necessary condition for aluminum smelting. The examples of Canada and Norway show that the production of primary aluminum can be based on imported bauxite (or alumina) and cheap hydroelectricity. In contrast, Jamaica, with plenty of bauxite, has not been a location for aluminum smelting.

Table 6-5. Bauxite Reserves in Brazil, 1981
(metric tons)

Region	Measured	Indicated	Inferred
North	2,538,800,368	2,041,581,330	416,289,558
Amapá	41,077,304	12,998,104	31,133,704
Amazonas	33,004,776	—	—
Pará	2,464,718,288	2,028,583,226	385,155,854
Southeast	71,645,115	12,814,740	5,041,522
Espírito Santo	1,091,908	—	—
Minas Gerais	66,457,108	12,220,932	4,527,811
Rio de Janeiro	2,135,793	241,181	29,315
São Paulo	1,960,306	352,627	484,396
South	4,917,779	58,884	—
Santa Catarina	4,917,779	58,884	—
Total	2,615,363,262	2,054,454,954	421,331,080

Source: Departamento Nacional da Producção Mineral (DNPM), *Anuário Mineral Brasileiro* (Brazilian Mineral Yearbook) (Brasília, DNPM, 1982) p. 122.

Table 6-6. Company Bauxite Reserves in the Amazon Region

Company	Location	Million metric tons	Percentage of total
CVRD	All locations	2,031	45
	Trombetas[a]	276	
	Almeirim	250	
	Paragominas	1,100	
	Paragominas[b]	360	
	Carajás	45	
Alcoa	All locations	700	15
	Trombetas[c]	300	
	Trombetas	400	
Rio Tinto Zinc	Paragominas[b]	640	14
CBA	All locations	310	7
	Trombetas[a]	60	
	Paragominas	250	
Shell	All locations	260	6
	Trombetas[c]	200	
	Trombetas[a]	60	
Ludwig	Almeirim	200	4
Reynolds	All locations	200	4
	Trombetas[a]	30	
	Juruti-Trombetas	170	
Alcan	Trombetas[a]	144	3
Enjex[d]	Almeirim	50	2
Norsk Hydro	Trombetas[a]	30	1
Total		4,565[e]	100[f]

Source: Paulo Sá, "Alumínio" (Aluminum). Unpublished paper (Brasília, Coordenação da Produção Mineral, CNPq, November 1981) p. 3.

[a] Mineração Rio do Norte: 46 percent CVRD, 24 percent Alcan, 10 percent CBA, 10 percent Shell, 5 percent Norsk Hydro, and 5 percent Reynolds.

[b] Mineração Vera Cruz (MVC): 64 percent Rio Tinto Zinc and 36 percent CVRD.

[c] Acquired from the Ludwig group.

[d] A Brazilian company.

[e] Measured and indicated. The difference between this figure and those of table 6-5 is due to rounding.

[f] May not sum to 100 due to rounding.

What, then, is the importance of bauxite to aluminum smelting? For Brazil, the existence of abundant bauxite has contributed to a favorable government policy toward aluminum smelting. Countries that are well endowed with bauxite are more likely to favor domestic refining and smelting and to design policies compatible with that goal. Those policies in turn may be a decisive factor in creating or magnifying international competitiveness in aluminum smelting.

Public Policy

The Brazilian government's objective of processing bauxite into aluminum was matched by the interest of primary producers elsewhere to find new sources to replace capacity they had closed. In pursuit of their goals, these producers were often willing to supply financial resources, technology, and marketing capability to new capacity projects located in LDCs.

Yet as the potential for aluminum smelting in the North became recognized, it also became apparent that further expansion of the industry in the Southeast was precluded by the unavailability of additional low-cost electricity. The shift in competitive advantage from the Southeast to the North is shown in figure 6-5 in which S_{SE} and S_N represent the long-run social marginal costs of electricity in the Southeast and the North. Costs initially were lower in the Southeast than in the North because of a developed infrastructure, abundant labor, and lower transmission costs. As the demand for electricity in the Southeast increased from D_{SE} *to* D'_{SE}, however, more costly electric power had to be developed. If marginal cost pricing were used, the price of electricity would increase from P_1 to P_2 as consumption increased from Q_1 to Q_2. The process of growth thus exhausted the cheap electricity available in the Southeast so that a point (A) was reached at which the development of electric power in the Southeast became more costly than in the North. This shift of advantage from region to region in Brazil reflects the same process of regional shifts that occurred within the United States (see chapter 2).

Because electricity could not be shipped from the North to the Southeast and demand in the Southeast continued to grow (D''_{SE}), higher cost sources of electric power continued to be developed in the Southeast. But once point A was reached, it became more advantageous in terms of cost to construct new generating capacity in the North. If that cost advantage could be reflected in the electricity rates, aluminum smelters would have an incentive to build capacity in the North.

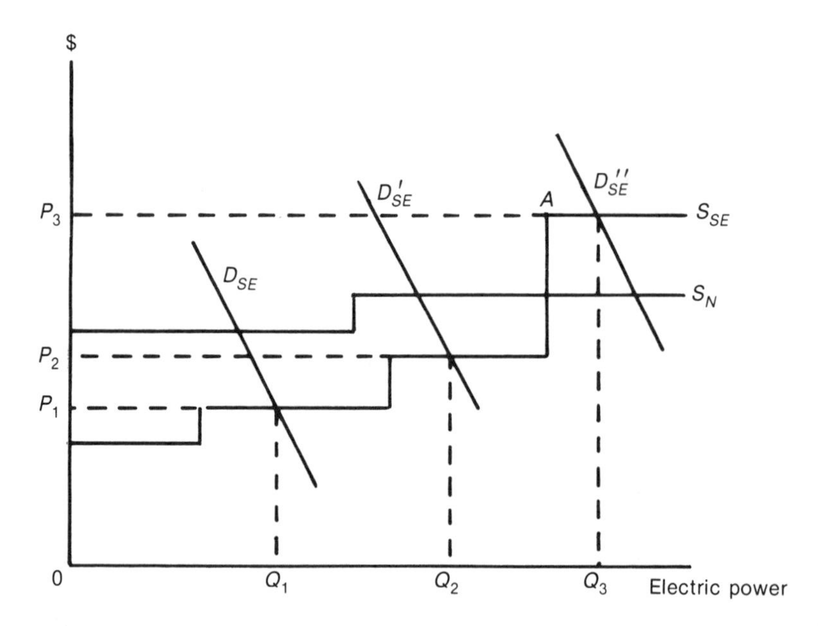

Figure 6-5. Shift in comparative advantage in Brazil from the Southeast to the North.

One characteristic of hydroelectric power in the north of Brazil is that it must be developed in large blocks because of economies of scale in hydro generation. Yet the demand for electricity in the region was very low; any new, large, available blocks of power would create a surplus. Two options were open to the government. The first was not to develop hydropower in the region, thus limiting its economic development. The second option was to develop hydropower, knowing that a surplus would be created. If demand for electricity was expected to grow slowly, the surplus would be available for a long time at a very low opportunity cost (only the operating cost of generation). One way to speed up the shifts in electricity demand would be to offer favorable rates to energy-intensive industries. Although the full social cost of the hydro plant would not be recovered, once the plant was in operation, all revenues above the variable costs would contribute to the payment of the fixed costs, which are incurred whether or not the plant operates. Of course, if the rate charged approaches the long-run social cost, the loss would be reduced. Still, that loss had to be compared to the social benefits resulting from the availability of electricity.

Aluminum smelting fits into this scenario well because it uses large quantities of electricity. It is also capital intensive, an advantage in a region in which labor is scarce. Reduced electricity rates for aluminum smelting might be a rational policy to accelerate the growth in electricity demand when such demand initially is low. The benefit offered to aluminum smelting would tend to be eliminated over time, however, as growth in demand for electricity eventually raised the opportunity cost of the electricity used for aluminum smelting.

Brazil's favorable government policy toward the development of aluminum smelting seems to have been based on these considerations. The development of the aluminum industry in the Amazon region, however, is part of a global strategy for regional development, especially the eastern portion. That strategy, "the Grand Carajás program," was initiated following the discovery of large iron ore deposits in the Serra dos Carajás in the state of Pará. To promote the iron ore project of Serra dos Carajás, CVRD prepared a proposal for global development of the natural resources of the Eastern Amazon region, based on a study done by the International Development Center of Japan.[15] The study proposed a massive, export-oriented plan for the development of natural resources, agriculture, and cattle raising over an area of 900,000 square kilometers. The development region included most of the state of Maranhão and portions of the states of Pará and Goiás. Investment for the program was estimated to be a total of $167 billion.[16] Of this total, $22.5 billion dollars was for infrastructure, which was to be provided by the government. Included were the 4,000-megawatt Tucuruí hydropower plant, a 900-kilometer railroad from Serra dos Carajás to São Luís, sea and river ports, roads, and urban centers. The mineral projects in the plan totaled $7.9 billion, of which $3.3 billion would be for alumina refining and aluminum smelting (table 6-7).

Besides providing infrastructure the government offered a series of incentives to industries locating within the Grand Carajás area. Those incentives, available to most firms, included a ten-year income tax holiday, exemption from taxes on imported equipment, and leasing of federal lands under favorable conditions.

For aluminum smelting a key consideration was the availability of low-cost electricity, which was to be provided by the Tucuruí hydropower project. The project's first phase, to be completed in 1988, will provide a capacity of 4,000 megawatts, of which 600 to 700 megawatts initially will be unused. (Because each phase is composed of various stages, however, the plant is already generating some electricity.) The total cost of the first phase is estimated to reach $3.8 billion, of which $300 million are transmission costs.[17] Financing in foreign currency for this plant comes from 28 sources

Table 6-7. The Carajás Program Investment in Mineral Projects (millions of dollars)

Project	Investment
Carajás (iron ore)	4,600[a]
Albrás	1,289
Alunorte	571
Alumar	1,475
Total	7,935

Source: Paulo Sá, "Grande Carajás: Tema para Reflexão" (Grand Carajás: Theme for Reflection), *Revista Brasileira de Tecnologia* vol. 13, no. 1 (January/March 1982) pp. 34 and 39.

[a] $2.8 billion for infrastructure.

at different interest rates that vary from 4.9 to 17.7 percent, although most of the rates are around 12 percent.[18]

This project guarantees smelters a supply of power at favorable rates. By an order of the Ministry of Mines and Energy, producers of primary aluminum in the Amazon region will be ensured a 20-year supply of electricity at rates below those paid by aluminum producers in southeastern Brazil. (For example, the discount is 15 percent for Albrás and 10 percent for Almar.) Rates for both projects are further limited by a provision that the total electricity cost in aluminum smelting will not exceed 20 percent of the price of aluminum in the international market. Further discounts may be granted to smelters located near the power plants or if such discounts are deemed necessary for the viability of specific projects. With these measures the government has ensured an advantage for smelters in the North compared with those in the Southeast and, more importantly, with those in most other countries. If the trend of rising electricity prices continues, the 20 percent limit on electricity costs will be a decisive advantage to aluminum producers in northern Brazil.

An example of the effect of the discounts in electricity rates is given by R. C. Machado.[19] In September 1979 the electricity rate in the Southeast was 14.30 mills/kWh. The 15 percent discount given Albrás would set the rate in the North at 12.57 mills. Considering the aluminum prices prevailing around 1980, this rate was well below the highest allowable rate set by the government—that is, 20 percent of the international price of primary aluminum. The full cost of electricity from Tucuruí, which had begun to supply aluminum smelters, was estimated at 16.3 mills/kWh; thus, the smelters were buying electricity at less than its full cost.[20] We have no estimates, however, of the long-run marginal cost of electricity for aluminum smelting.

Public policy in northern Brazil has undoubtedly favored export-oriented mineral projects and especially primary aluminum production. The advantage of a large, unexploited hydropower potential was reenforced by a set of favorable policy measures. The success of those measures is shown by the two projects Alumar and Albrás, which are currently under construction.

Yet the policies adopted by the government have faced strong criticism in regard to the costs and benefits for Brazilian society of the Grand Carajás program. It has been argued that the export-oriented program, conceived as a source of foreign revenue, will neither provide those revenues on a significant scale nor be the best option for development of the Amazon region. It is also considered to be beyond Brazil's capacity to raise the financial resources needed to carry out the program. Further concern has arisen regarding the effects of the program on the ecosystem of the Amazon region and its impact on local populations, including Indian communities.

Specifically, in terms of the aluminum industry, the policy question is to what extent the subsidies, fiscal incentives, and investment in infrastructure equal or exceed the benefits to society as a whole. In addition, it is not clear whether such extensive incentives were really needed. The advantages given to Albrás seem excessive when compared to what was offered to Alumar. Yet major changes in public policy are not likely to occur in the near future. From the viewpoint of primary producers, there will continue to be an advantage in locating aluminum smelting capacity in northern Brazil.

CONCLUSION

Brazil became an exporter of primary aluminum with the entry of Valesul into the industry in 1982, and it has improved that position with the development of Alumar and Albrás. With the recent start-up of these two smelters, the aluminum industry in Brazil now has two different sectors, one in the Southeast and one in the North. The Southeast was developed to serve domestic demand, based on a combination of local inputs and proximity to consumers. The industry grew under government protection to reach an efficient scale and in the process became internationally competitive. Yet the same overall economic growth that increased aluminum demand was also responsible for an increased demand for electricity, which in turn increased the opportunity cost of electricity. The result is that the aluminum industry in the Southeast is competitive but only for its existing capacity. It is not competitive with the North or with other countries for the construction of new capacity.

Alcoa and Alcan do not intend to expand their operations in the state of Minas Gerais because of limitations on bauxite resources, as well as the unavailability of low-cost electricity in the most industrialized region of Brazil. CBA, which generates most of its own electricity and has better bauxite deposits, still plans to add capacity in the Southeast. Alunordeste (Alcan's smelter in the state of Bahia) and Valesul may increase capacity, but their plans are uncertain. Overall, significant growth of aluminum smelting in the Southeast is highly unlikely.

The industry in the North, in contrast, has a large hydropower potential and is favored by government policies as part of an overall strategy to develop the natural resources of the region. The main measures to encourage development of the aluminum industry were the guarantee of a 20-year supply of electricity at rates 15 percent below those paid by smelters in the Southeast and a limit on the total cost of electricity to 20 percent of the price of aluminum in the international market. With these measures the government provided an advantage that ensured that aluminum produced in the Amazon region would be competitive in the international market.

The development of aluminum smelting in the Amazon region has created a new export-oriented sector in the Brazilian aluminum industry. Alumar started operations in 1984 and will reach an annual capacity of 350,000 tons of primary aluminum. Albrás started operations in 1985, and its capacity is planned eventually to reach 320,000 tons.

Exports of primary aluminum from northern Brazil will be ensured by existing contracts. There could be controversy within Brazil, however, if domestic demand again exceeds domestic supply, as historically has been the case. Importing primary aluminum at world prices while exports are sold at a discount would be a very sensitive issue.

The level of domestic demand is a basic variable in assessing the impact of Brazilian production in the international market. Although it seems certain that Brazil will be an important producer of aluminum, it is not so certain that its position as an exporter of primary metal will be as significant as the original plans indicated.

It seems likely that the aluminum sector would have developed in the Southeast by the early 1970s even without the protection that existed in that decade. The size of the market justified production at an efficient scale, and the inputs, particularly electricity, were competitively priced. If the southern industry had not been well established, policy for other scarce mineral commodities suggests there might have been two differences in the development of the industry in the North: (1) the priority would be the supply of the domestic market, and (2) the participation of the Brazilian government in the industry would have been even greater to ensure that imports were reduced.

According to recent estimates, primary aluminum capacity in Brazil is expected to reach between 1.25 million and 1.3 million tons in 1995.[21] Should that occur, capacity will have increased by about 900,000 tons during the period 1982–1995, which is likely to make Brazil an important factor in the world aluminum market. Almost all additions to capacity will be in the two projects of Alumar and Albrás, both in northern Brazil.

NOTES

1. The author appreciated the very helpful comments and suggestions made by Merton J. Peck, John E. Tilton, and Richard L. Gordon. He is also grateful to Raymundo C. Machado for the valuable information he provided.
2. Independent fabricators in Brazil are rather small. Late in the 1970s, they were using about 80 percent of all imports and were buying the imported metal at 50 percent above the international price because of compulsory deposits and taxes (see Marcos Dantas, "A Questão do Alumínio—O Presente e o Futuro da Indústria de Alumínio no Brasil" (The Aluminum Question—Present and Future of the Aluminum Industry in Brazil). Paper presented to the I Congresso de Defesa da Amazônia, Ouro Preto, Minas Gerais, September 1980). The dependence of the fabricators on imports ended with the start-up of Valesul in 1982.
3. Dantas, "A Questão do Alumínio," p. 24.
4. Orlando Euler de Castro, "A Indústria e o Mercado de Alumínio no Brasil" (The Brazilian Aluminum Industry and Market). Paper presented to the Simpósio sobre el Alumínio en Latinoamerica, Organización de los Estados Americanos, Oaxtepec, Mexico, March 1983, p. 88.
5. Ibid.
6. Ministério das Minas e Energia, *Modelo Energético Brasileiro* (Brazilian Energy Model) (Rio de Janeiro, Gráfica e Editora Celsus Ltda., 1981) p. 91.
7. Ibid., p. 17.
8. Gerald M. Meier, *Leading Issues in Economic Development*, 3rd ed. (New York, Oxford University Press, 1976) pp. 652–653.

9. In 1981 Alcoa was granted an exclusive permit to import aluminum and to sell it together with its domestic production at a price that was the weighted average of the import and domestic prices.
10. The company, which was called COMEC, was created by CVRD in 1976. It was abolished in 1979.
11. Raymundo C. Machado, *Alumínio Primário no Brasil* (Primary Aluminum in Brazil) (Ouro Preto, Minas Gerais, Fundação Gorceix, 1982) p. 96.
12. Paulo Sá, "Grande Carajás: Tema para Reflexão" (Grand Carajás: Theme for Reflection), *Revista Brasileira de Tecnologia* vol. 13, no. 1 (January/March 1982) p. 38.
13. Paulo Sá, "A CVRD e a Indústria de Alumínio" (CVRD and the Aluminum Industry) (Brasília, Conselho Nacional de Desenvolvimento Científico e Tecnológico—CNPq, April 1981) p. 47.
14. Sá, "Grande Carajás," pp. 38–39.
15. Ibid., p. 31.
16. Argemiro Ferreira, "Carajás—O Grande Desafio" (Carajás—The Great Challenge), *Ciência Hoje* vol. 1, no. 3 (November/December 1982) pp. 31–34.
17. Eletronorte, "UHE Tucuruí" (The Tucuruí Hydropower Plant) (1983) p. 2.
18. Eletronorte, "Demonstrativo dos Financiamentos e Empréstimos a Pagar para os Exercícios Findos em 31 de Dezembro" (Statement of Outstanding Credits and Loans at Yearend) (1983).
19. Machado, *Alumínio Primário no Brasil,* p. 106.
20. Ibid., p. 93.
21. Carlos Romano Ramos, oral communication, August 1987.

7

CANADA: AN EXPANDING INDUSTRY

CARMINE NAPPI

Canada is a clear winner in the restructuring of the world aluminum industry: by 1985, it was second only to the United States in aluminum production. Its smelting capacity of 1.29 million tons is 8.6 percent of the total capacity in market economies. Moreover, because Canada consumes only 300,000 tons annually, it is the world's largest exporter of primary aluminum.

The healthy state of the Canadian aluminum industry is often attributed to the availability of low-cost hydropower. Other factors include a location with easy access to the large American market; the stability of public institutions; and the quality of the nation's economic and social infrastructures. Although these factors have been present since the turn of the century, however, the Canadian aluminum industry until recently has remained relatively small. In the 1960s and 1970s, for instance, there was no expansion of capacity. The smelter at Grande-Baie in Quebec, which began operating in 1980, was the first new smelter in Canada in over twenty-five years—a period in which Canada's share of Western aluminum production capacity declined from 20.5 percent to 8.6 percent.

For Canada, the 1980s have been the decade of expansion. Once the Grande-Baie plant reaches normal operational capacity, Alcan will be operating seven smelters with a total capacity of over 1.1

Carmine Nappi is professor of mineral economics at the École des Hautes Études Commerciales at the University of Montreal.

million tons per year. In addition, Reynolds has recently increased its production capacity at the Baie-Comeau, Quebec, smelter from 154,000 tons, to 279,000 tons; Pechiney has just finished work on a 230,000-ton smelter at Bécancour, Quebec. Finally, feasibility and profitability studies are being carried on by Alcan at Laterrière, Quebec, and Vanderhoof, British Columbia; by Kaiser in Sept-Iles, Quebec, with the possible addition of Alusuisse; and by Alcoa in Manitoba. All of the big six thus are active in Canada.

The completion of these projects will more than double the country's production capacity. Such a high rate of expansion suggests that Canada is now an extremely favorable location for aluminum smelting. This chapter examines why the comparative advantage of Canada in aluminum smelting has changed in recent years and how public policies have helped or hindered the country's industry in adjusting to its new competitive environment.

The first section, a historical perspective, will discuss the main features of the Canadian aluminum industry and sketch the major stages in its development up to the early 1980s. The second section will study Canada's comparative advantages in the production of primary aluminum. Although alumina, capital, and electricity costs account for more than 70 percent of total production costs, we will focus on the cost of electricity because, as chapter 1 indicates, this is the input whose price varies the most among countries. Our evaluation will be made in the context of the major structural changes that have occurred in the international aluminum industry over the last ten years: the increase in energy costs as a proportion of production costs; the reduction in the degree of concentration in the industry with the entry of private or government producers; the rise of substitute products; and the increased importance of recycling and maturing markets. Finally, this section examines to what extent the construction of new smelters corresponds to the shift in Canada's comparative advantages in the production of primary aluminum.

Canada has not developed economic policies specifically directed at the aluminum industry, although the industry does feel the effects of federal and provincial policies on employment, taxation, environment, energy, foreign investment, or regional development. The last section of the chapter will examine some of these policies to determine whether they have facilitated or hindered the adjustment of Canadian producers to the changes that have occurred in the structure of the international aluminum industry over the last ten years.

THE DEVELOPMENT AND MAIN FEATURES OF THE CANADIAN ALUMINUM INDUSTRY

Major Development Stages

The development of an aluminum industry in Canada[1] took place in four major stages (table 7-1). The first stage (1899–1930) comprises the setting up and consolidation of Alcoa in the Shawinigan region of Quebec to take advantage of the abundance of hydroelectric power and to play a role in international cartels through its Canadian subsidiary (such activities were prohibited for Alcoa by the Sherman Act). This period also witnessed the shift in 1922 of production from the Shawinigan area to the Saguenay-Lac-Saint-Jean region, where 75 percent of Alcan's Canadian aluminum production capacity remains today. It was during this period that Alcoa set up a Canadian company (Aluminium Limited) to oversee all of its foreign operations except the bauxite mines in Surinam. Alcan was a wholly owned subsidiary of Alcoa until 1928, and it then became an independent company with, however, the same major shareholders as Alcoa. (Alcan was created by distributing its stock to Alcoa shareholders.)

The second stage (1939 to the mid-1950s) was marked by two important events. The first was World War II, during which Aluminium Limited (later renamed Alcan), with the assistance of the British and U.S. governments, expanded its production capacity. Alcan has always owned most of the hydropower that it uses, and during this period the installed capacity of the company's power stations was increased by 896,000 kilowatts with the building of the Shipshaw station, bringing Alcan's installed capacity in the Saguenay-Lac-Saint-Jean region over the 1.5 million kilowatt mark. This period was extremely important for Alcan: the war transformed a marginal aluminum producer into a giant. The second event was the decision of Judge John C. Knox, after Alcoa had been found guilty of Sherman Act violations, to require Alcoa shareholders to divest themselves of their Alcan or Alcoa shares over a ten-year period; the common ownership of the two companies was held to limit competition.

The third stage (1958–1980) reflects Alcan's strategy in those years. Until the mid-1950s, Alcan had considered itself a supplier of primary aluminum for American and overseas markets. Unlike other North American producers, Alcan had not invested in fabrication plants. In 1958 it decided to make up for lost time in the area of

Table 7-1. Chronology of the Aluminum Industry in Canada

August 14, 1899	Contract signed by the Pittsburgh Reduction Company (later Alcoa) and the Shawinigan Water and Power Company for the purchase of hydraulic power at Shawinigan, Quebec.
October 18, 1901	End of construction work on aluminum plant, and power station started on May 23, 1900; facilities are put into service.[a]
July 3, 1902	Northern Aluminum Company Limited established as a wholly owned subsidiary of Alcoa to direct the American company's new Canadian assets.
1906–1922	Production capacity of Northern Aluminum's plant doubled. Wire and cable plant and specialized lamination and molding plant built.
1922–1926	First power station built at Isle-Maligne in the Saguenay-Lac-Saint-Jean region by the Saguenay Power Company, later bought by Alcoa. The American aluminum company also acquires port facilities at Port Alfred and two railway companies. Smelter built in 1925 at Arvida, 20 miles from Port Alfred.
May 31, 1928	Alcoa is reorganized and Aluminum Limited established in Canada, with responsibility for the assets of all Alcoa's overseas operations, including Northern Aluminum but excluding interests in bauxite mines in Surinam.[b]
1928–1937	Aluminum Limited is reorganized and becomes an international company. Swiss office opened; regional Canadian office (later headquarters) set up in Montreal; regional management of interests in British Guiana (Demerara Bauxite Company), England, India, Italy, Norway, and the Far East.
1939–1945	With increased industrial and military demand for aluminum, Aluminum Limited undertakes a major expansion program: Canadian production capacity (100,000 tons in 1939) increases fivefold over four years. Arvida plant is enlarged, Beauharnois plant is built in 1942, and Isle-Maligne plant in 1943; profits up 600 percent; with building of Shipshaw power station, installed capacity increases by 896,000 kW.
1946–1954	As military demand disappears, Aluminum Limited develops and enlarges the postwar international civilian market for aluminum. During this period, demand outstrips industrial production, thanks to a price fall made possible by improved production techniques and economies of scale. Two more stations are built in the Saquenay-Lac-Saint-Jean region on the Peribonka River: the Chûtes-du-Diable and Chûte-à-la-Savane stations.
July 6, 1950	After thirteen years of legal proceedings, Judge John C. Knox concludes that it is doubtful whether competition as to prices and efficiency exists to the same extent as it would if Alcoa

Table 7-1 (continued)

	and Aluminum Limited were totally separate. To benefit the public, he directs Alcoa shareholders to divest themselves of their Aluminum Limited or Alcoa shares over a ten-year period.
August 1954	On December 30, 1950, Aluminum Limited signs agreement with the British Columbian government for the construction of a smelter, electric power complex, and town at the mouth of the Kitimat River. The first aluminum ingot is poured four years later in August 1954. Conveniently located near abundant power supplies, the complex is planned to penetrate the important Western U.S. and Japanese markets. Originally supplied with alumina from an Alcan plant in Jamaica, today Kitimat mainly uses Australian alumina.
1955	A second aluminum company opens for business in Canada: Canadian British Aluminum Co. Ltd., formed by British Aluminum (64 percent) and O.N.S. Paper Co. Ltd. (36 percent) at Baie-Comeau, Quebec. Reynolds and Tube Investments acquire British Aluminum's interest in 1958 and O.N.S. Paper's in 1966. In 1970 the 154,000 tons/year plant becomes the exclusive property of Reynolds and is renamed Canadian Reynolds Metals Company Ltd.
1958–1978	In its annual report for 1958, Aluminum Limited announces that it will henceforth pour more resources into building and enlarging fabricating plants to increase sales of aluminum products and find new markets for its primary aluminum. Additions to fixed assets for fabrication rise from $58 million for 1951–1959 to $756 million for 1960–1973. Aluminum Limited builds (1956–1960) a sixth power station at Chûtes-des-Passes with an installed capacity of 750,000 kW. In 1966, Aluminum Limited changes its name, becoming Alan Aluminum Limited, with its principal subsidiary Aluminum of Canada Limited.
1978–1982	Alcan Aluminum builds the Grande-Baie, Quebec, electrolysis plant with a current capacity of 171,000 tons.
July 1983	An agreement is signed by the Quebec provincial and French governments for the construction at Bécancour of a 230,000 tons/year smelter. Pechiney on the French side (66.7 percent) and the Société Générale de Financement on the Quebec side (33.3 percent) are responsible for the project. On January 27, 1984, the board of directors of Alumax votes to join Pechiney and the Quebec provincial government as partners in the project. Ownership and responsibility are now shared as follows: Pechiney (51.1 percent), Alumax (24.95 percent), and Société Générale de Financement (24.95 percent). The smelter started operations on April 20, 1986.

Continued

Table 7-1 (continued)

April 10, 1984	Alcan announces the construction of a new 248,000 tons/year primary aluminum smelter at Laterrière, Quebec. The construction will proceed in three 82,700 tons/year phases spanning seven years. The first phase is planned to come onstream in mid-1988. When the project is complete, 133,000 tons/year of capacity will be idled at the aging 438,000 tons/year Arvida, Quebec, smelter. The remaining 115,000 tons/year of new capacity at Laterrière will be a net addition to Alcan's worldwide production capacity.
May 1985	Because of depressed aluminum market, Alcan announces that the Laterrière aluminum smelter has been postponed.

Source: Derived from information contained in the following: I. A. Litvak and C. J. Maule, *Alcan Aluminum Limited: A Case Study* (Ottawa, Ministry of Supply and Services, February 1977); Royal Commission on Corporate Concentration, "Case Study No. 13," chapters 1, 2, and 3; and *Metals Week,* various issues.

[a] On December 2, 1901, 67,200 pounds of aluminum leaves Shawinigan bound for Yokohama, Japan. Canada is already becoming a major aluminum exporter.

[b] Alcoa also transfers its interests in bauxite reserves in British Guiana to Aluminum Limited because mine leases stipulate that bauxite must be transformed on British soil.

fabricating. Whereas 21 percent of its additions to fixed assets went to fabrication between 1951 and 1966, that figure rose to 49 percent for the 1967–1972 period. The 1950s also witnessed the construction of the first Canadian smelter outside Quebec at Kitimat, British Columbia, and the entry of Reynolds with its smelter at Baie Comeau, Quebec. There were virtually no additions to existing capacity in the following two decades (the exception being the new Grande-Baie smelter). As a result, Canada's share of world capacity fell from 22.5 percent to 8.5 percent.

In the mid-1980s we may well be on the threshold of a fourth stage, which will see an increase in the number of producers attracted to Canada by abundant, stable, and relatively inexpensive electricity. The subsequent discussion focuses on the 1980s for an analysis of Canada's position in the world aluminum industry.

The Structure of the Canadian Aluminum Industry

The Canadian aluminum industry of the 1980s is the legacy of the developments discussed above. Figure 7-1 presents a diagram of the industry for the year 1980. The following features need to be underlined.

First, the Canadian aluminum industry has been oriented toward exports from its inception. As Paul Clark puts it:

> Canada's position as the world's principal exporter of aluminum began at an impressive level. On December 2, 1901, the substantial amount of 67,200 pounds of aluminum ingot left the Shawinigan platform crated for delivery to Yokohama, Japan. . . . The big shipments headed overseas where Shawinigan aluminum became a familiar sight on the docks of the World, especially Liverpool, Rotterdam, Trieste and Hamburg.[2]

Table 7-2 shows that this export orientation has remained the case in recent years. During the 1960s exports represented more than 80 percent of Canadian aluminum production; in the 1970s, exports to foreign markets averaged 72 percent of production. Table 7-2 also shows that Canadian export and production rates, which are closely linked, move in a pattern that is different from domestic consumption.

In the previous section, we stressed that from 1958 on, Alcan greatly increased its emphasis on fabrication. Although a growing percentage of the funds invested in fabrication was spent in Canada, most of the investment in fabricating plants was in the United States. Table 7-2 traces the increasing integration of the Canadian aluminum industry to the American market resulting from this trend. The table shows that the percentage of primary aluminum exported to the United States jumped from 18 percent in 1960 to 65 percent in 1985. Thus, although Canada has remained the leading international exporter of aluminum, it has concentrated its exports in the U.S. market. Such a shift has served to supply Alcan's vertically integrated U.S. fabricating plants, facilities that were located in the United States because of transport costs and a higher U.S. tariff on fabricated products.[3] It also reflects the fact that Canadian exports were squeezed out of the growing Japanese and European markets by the expansion of domestic capacity in those regions. Given the properties and characteristics of aluminum (an extremely versatile engineering and construction material and a highly flexible and potentially energy-efficient metal), these countries preferred to replace imports with domestic production, even when protective measures might be required to attain this objective. Finally, we should note that, although American imports from Canada represent on average less than 10 percent of U.S. consumption of primary aluminum, they are a healthy 65 to 80 percent of total U.S. aluminum imports.

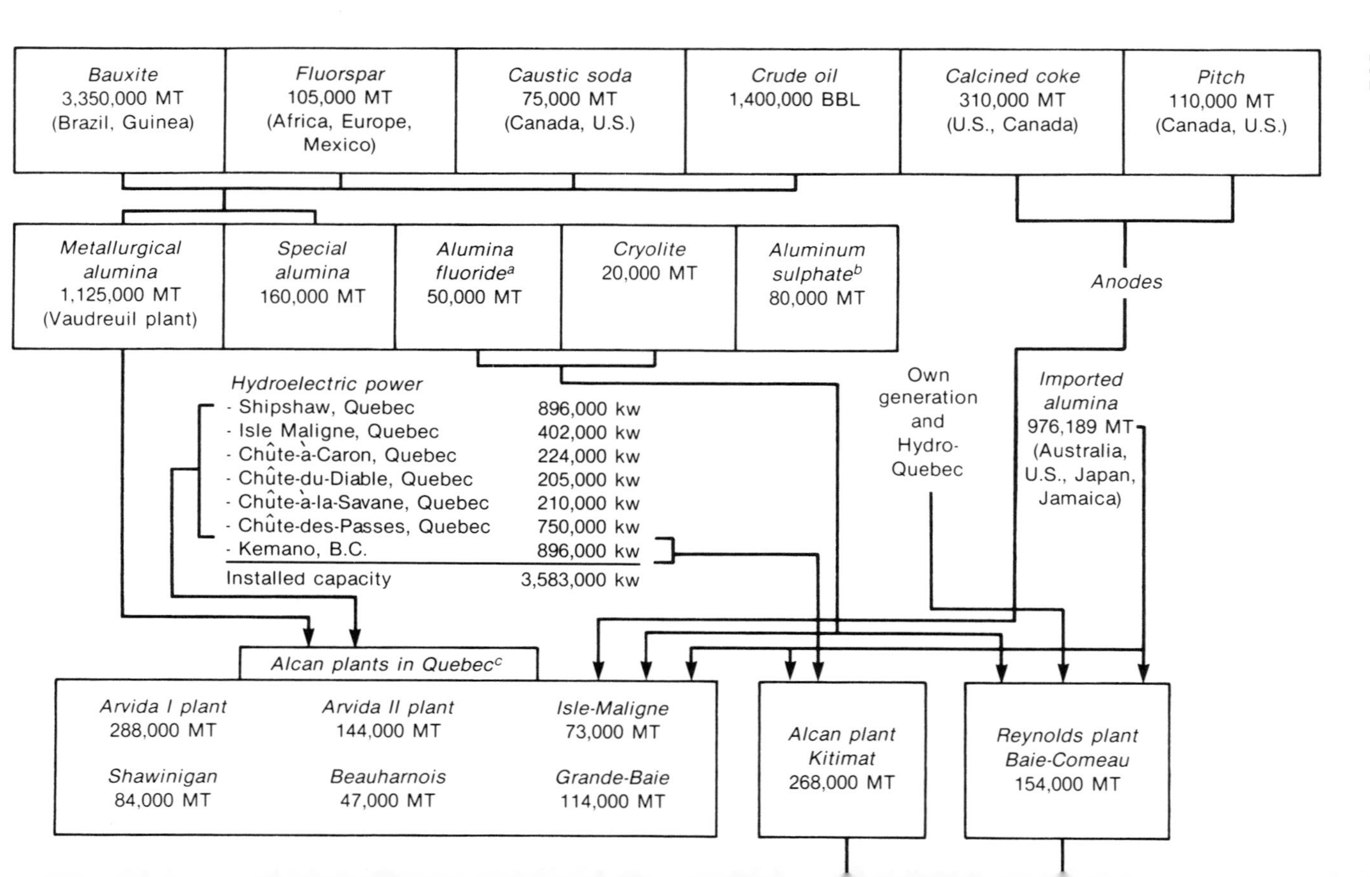
Bauxite
3,350,000 MT
(Brazil, Guinea)
Fluorspar
105,000 MT
(Africa, Europe, Mexico)
Caustic soda
75,000 MT
(Canada, U.S.)
Crude oil
1,400,000 BBL
Calcined coke
310,000 MT
(U.S., Canada)
Pitch
110,000 MT
(Canada, U.S.)
Metallurgical alumina
1,125,000 MT
(Vaudreuil plant)
Special alumina
160,000 MT
Alumina fluoride[a]
50,000 MT
Cryolite
20,000 MT
Aluminum sulphate[b]
80,000 MT
Anodes
Hydroelectric power
- Shipshaw, Quebec 896,000 kw
- Isle Maligne, Quebec 402,000 kw
- Chûte-à-Caron, Quebec 224,000 kw
- Chûte-du-Diable, Quebec 205,000 kw
- Chûte-à-la-Savane, Quebec 210,000 kw
- Chûte-des-Passes, Quebec 750,000 kw
- Kemano, B.C. 896,000 kw
Installed capacity 3,583,000 kw
Own generation and Hydro-Quebec
Imported alumina
976,189 MT
(Australia, U.S., Japan, Jamaica)
Alcan plants in Quebec[c]
Arvida I plant
288,000 MT
Arvida II plant
144,000 MT
Isle-Maligne
73,000 MT
Shawinigan
84,000 MT
Beauharnois
47,000 MT
Grande-Baie
114,000 MT
Alcan plant
Kitimat
268,000 MT
Reynolds plant
Baie-Comeau
154,000 MT

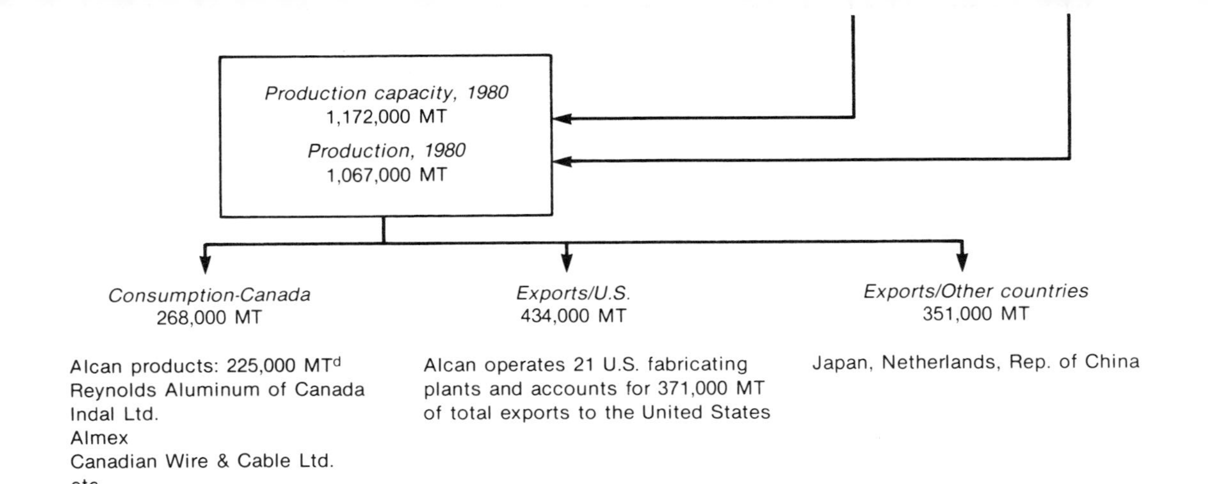

Figure 7-1. Diagram of the Canadian aluminum industry, 1980. *Sources:* Metal Bulletin, *World Aluminum Survey,* 5th ed. (London, 1981); Alcan, *Alcan Facts 1962* (Montreal, Alcan Aluminum Limited); and various public documents from Alcan Aluminum Limited.

[a]This product is derived from fluorspar, but the arrow was omitted to simplify the diagram.

[b]All of the aluminum sulphate produced in Alcan's chemical plants is sold, 95 percent of it in Quebec where it is used by water purification plants and the pulp and paper industry.

[c]Although Alcan produced 20,000 MT of cryolite in 1980, only 5,000 MT were used in its electrolysis plants; the difference was sold, notably to Reynolds of Canada. The same holds true for aluminum fluoride, of which only half is used in Quebec plants.

[d]Alcan products fabricating plants are largely specialized in the production of wire and cable (Lapointe, Saint-Augustin, Shawinigan, and Bracebridge plants); sheet and foil (Sanguenay, Kingston, Victoriaville); extrusions (Laval, Aurora, Winnipeg, Calgary); building products, and so forth.

Table 7-2. Canadian Production, Consumption, and Exports of Primary Aluminum, 1960–1982

	Production		Domestic consumption		Exports					
					Total			To United States		
Year	Volume[a]	AGR[b]	Volume	AGR	Volume	AGR	PTE[c]	Volume	PEUS[d]	PEI[e]
1960	691[f]	—	101	—	501	—	73	91	18	65
1961	601	−13	115	14	442	−12	73	107	24	60
1962	625	4	134	17	523	18	84	192	37	70
1963	652	4	138	3	576	10	88	249	43	66
1964	764	17	156	13	570	−1	75	231	41	65
1965	753	−1	169	8	642	13	85	316	49	67
1966	807	7	190	12	650	1	81	347	53	74
1967	884	10	181	5	690	6	78	329	48	81
1968	888	0	209	15	783	14	88	437	56	71
1969	996	12	196	−6	804	3	81	371	46	88
1970	964	−3	220	12	762	−5	79	295	39	93
1971	1,016	5	258	17	805	6	79	408	51	82
1972	918	−10	279	8	691	−14	75	442	64	75
1973	941	3	301	8	699	1	74	384	55	83

1974	1,020	8	358	19	682	−2	67	367	54	80
1975	878	−14	286	−20	509	−25	58	309	61	74
1976	633	−29	322	13	508	0	80	356	70	70
1977	973	53	332	3	655	29	67	450	69	75
1978	1,048	8	339	2	863	32	82	471	55	69
1979	860	−18	340	0	551	−36	64	331	60	64
1980	1,067	24	292	−14	785	42	73	437	56	84
1981	1,109	4	268	−8	720	−8	65	484	67	76
1982	1,064	−4	274	2	896	24	84	419	47	68
1983	1,085	2	338	23	925	3	85	581	63	78
1984	1,215	12	412	22	832	−10	68	611	73	69
1985	1,275	5	—	—	1,050	26	82	683	65	79

Sources: The Aluminum Association, *Aluminum Statistical Review* (Washington, D.C., various years); A. L. Dorr and J. E. Tilton, "Bauxite and Aluminum," in C. E. Beigie and A. O. Hero, Jr., eds., *Natural Resources in U.S.-Canadian Relations* vol. 2, chap. 7 (1980) p. 227; and World Bureau of Metal Statistics, *World Metal Statistics* (London, various issues).

[a] In thousands of metric tons.

[b] Annual growth rate (AGR) as a percentage.

[c] Percentage of total exports in aluminum production (PTE).

[d] Percentage of exports to the United States in total exports (PEUS).

[e] Percentage of exports to the United States in total U.S. imports of aluminum ingots (PEI).

[f] Aluminum production does not necessarily equal consumption plus exports because imports must also be taken into account, as well as secondary aluminum and inventory variations.

Second, unlike Australia and Brazil, the Canadian aluminum industry is completely dependent on bauxite imports. Although Canada is endowed with clay and other ore containing a certain level of alumina, it is totally lacking in the ore that has the highest level of all: bauxite. Alcan's answer to this peculiarity in economic geography was to build the Vaudreuil alumina plant in the Saguenay-Lac-Saint-Jean region to refine imported bauxite. The plant's annual capacity of 1.3 million tons is almost sufficient to fill the needs of Alcan's Quebec smelters.[4]

Alcan's British Columbia smelter and Reynolds's Quebec plant are mainly supplied with imported alumina. The Reynolds smelter is supplied by the company's alumina plant in Corpus Christi, Texas (it provides about 30 percent of Canadian alumina imports); the alumina for Alcan's smelter at Kitimat, British Columbia, comes mainly from Queensland Alumina Limited in Australia, a company in which Alcan has a 21.4 percent interest.[5] Of course, bauxite and alumina are not the only raw materials needed for aluminum production. Other inputs are either produced locally (for example, cryolite and aluminum fluoride) or imported (fluorspar from Africa, Europe, and Mexico; caustic soda and coke from the United States; and so forth).

Third, our discussion of the origin of different input supplies leads to another basic feature of the Canadian aluminum industry: the leading producer owns hydroelectric generating facilities with a total installed capacity of 3.58 million kilowatts, of which 2.7 million kilowatts are firm power capacity. The ownership of such power sources brings security of supply for there is no risk of a utility breaking a long-term power contract. The energy needs of the Reynolds plant at Baie-Comeau are provided by plants it owns and by a long-term contract with Hydro-Quebec at a rate approaching that charged to major power users. In the next section, we will return to the subject of relative energy costs and the advantages enjoyed by Canadian producers.

Fourth, downstream Canadian aluminum operations are quite diversified, involving the production of a wide range of sheet products, extruded products, electrical conductors, irrigation and gas distribution pipe and tubing, fuel tanks, foil containers, and even boats. These products are intended mainly for the local market. In 1985, shipments of noningot aluminum products to the United States were a bare 6 percent of total Canadian-U.S. shipments, or a little over 50,000 tons (for primary aluminum, the volume was 675,000 tons). These U.S.-bound shipments are mainly aluminum sheet metal and constitute a scant 12 percent of American imports of semimanufactured aluminum products. But the American market still remains a main outlet for Canadian products; in 1985 it accounted for 68 percent of Canadian exports of semifinished aluminum-based products.

CANADA'S COMPARATIVE ADVANTAGES IN ALUMINUM PRODUCTION

The international aluminum industry has undergone major changes in the course of the last decade, of which the phenomenal rise in energy prices is undoubtedly the most important. At the start of the 1970s, the energy input represented an average of 15 percent of the total production cost of aluminum; the percentage has since doubled, and in the case of some plants, the figure is approaching 35 percent.[6] This rise alone explains a large part of the increase in production costs.

The cost structure of aluminum smelting is shown in table 7-3, which includes the production costs for a new plant (that is, "greenfield" investment) located in Australia or in the United States, compared to those incurred by an existing American plant.[7] For each of these broad categories the table identifies the relative share of different elements in widely varying estimates of total costs given by international institutions or by consulting firms.[8] The following observations are based on this information.

First, excluding the Springborn and Townsend study because of its particular treatment of depreciation and interest costs, the costs of alumina, electricity, and capital alone account for 70 percent of the production costs for a new smelter. This figure rises to 75 percent or more in the case of an existing plant. Other cost factors are marginal.

Second, the total production costs shown in table 7-3 do not take aluminum transport costs into account because the calculations were on an FOB basis. Given that some smelters are located at the very center of a large consumer market whereas others are in remote areas to take advantage of energy cost savings or large bauxite deposits, these costs must be considered in comparing the competitive position of aluminum plants in different countries. A study by the Organisation for Economic Co-operation and Development (OECD)[9] maintains that smelters located within the main consuming markets might have a freight cost advantage equivalent to 2 or 3

Table 7-3. Aluminum Production Cost Structure for a New Australian Plant, a New Integrated American Plant, and an Existing American Plant, 1980
(percentage)

	New Australian plant							New integrated U.S. plant	Existing U.S. plant	
Cost factors[a]	Springborn-Townsend	OECD	Anthony Bird	Brown	Hashimoto	Spector	Chase-I	Chase-II	Springborn-Townsend	Woods-Burrows[i]
Alumina	.218	.273	.243	.324	.269	.286	.272	.277	.276	.338
(bauxite)[b]	(.029)							(.089)	(.101)	(.091)
Electricity[c]	.153	.172	.174	.155	.152	.161	.202	.270	.284	.202
Coal and pitch[d]	.041		.077	.078					.048	.058
		.104			.088	.100	.070	.069		
Other raw materials[d]	.026		.015	.039					.029	.023
Direct labor	.035								.053	
		.068	.055	.055	.074	.097	.074	.040		.065
Supervision[e]	.005								.008	
Maintenance[f]	.047		.086						.023	.038
		.149		.126	.121	.055	.094	—		
General overhead	.028		.100						.033	.051
Insurance and local taxes	.026	—	—	—	—	—	—	—	.017	.031
Capital costs (depreciation and interest cost)[g]	.422[h]	.234	.250	.224	.296	.301	.288	.327	.228[h]	.184
Total production costs	1.000	1.000	1.000	1.000	1.000	1.000	1.000	1.000	1.000	1.000

Sources: Anthony Bird Associates, "Aluminum Analysis," *Metal Analysis and Outlook* vol. 13 (April 1982); M. Brown, A. Dammert, A. Meeraus, and A. Stoutjesdijk, *Worldwide Investment Analysis: The Case of Aluminum,* World Bank Staff Working Paper no. 603 (Washington, D.C., World Bank, July 1983) pp. 32–40; Chase Econometrics (Chase-I), reported in T. G. Langton, "Economic Aspects of the

Bauxite/Aluminum Industry," *Journal of Metals* (August 1980) p. 15; Chase Econometrics (Chase-II), *World Aluminum: Retrenching and Restructuring* (Bala Cynwyd, Pennsylvania, February 1982) pp. AL-100 to AL-132; H. Hashimoto, *Bauxite Processing in Developing Countries* (Washington, D.C., World Bank, January 1982); Organisation for Economic Co-operation and Development, *Aluminum Industry: Energy Aspects of Structural Change* (Paris, OECD, 1983) p. 44; Spector as reported in OECD, *Aluminum Industry*, p. 44; Springborn Laboratories and Phillip Townsend Associates, Inc., "The Impact of Energy Costs, Technological Change and Capital Equipment Costs Upon Raw Materials Competition: 1980–1985–1990" (Enfield, Conn., and Houston, Texas, undated) chap. 8; and D. W. Woods and J. C. Burrows, *The World Aluminum-Bauxite Market: Policy Implications for the United States*, Charles River Associates Research Report (New York, Praeger, 1980) tables 2-6 and 2-8.

[a] Total production costs differ considerably from study to study. For a new Australian plant, costs (all in U.S. dollars) range from $1,175/ton (Spector) to $1,740/ton of aluminum. Spector's study is alone in giving such a low total cost; totals estimated by other studies in most cases top $1,500/ton. Total estimated production costs for an existing plant are lower than for a "greenfield" project, averaging $1,350/ton.

[b] We have assumed here that it takes 1.94 short tons (ST) of alumina to produce one ton of aluminum. Studies differ on the bauxite-alumina ratio. Australian plants use domestic bauxite with a lower aluminum oxide content, using 2.8 ST per ton. Existing American alumina plants use just 2.15 ST of bauxite imported from the Caribbean according to Springborn-Townsend or 2.3 ST according to Woods-Burrows.

[c] The studies also differ considerably on the cost of electricity. Differences occur not only in the electricity consumption given per ton of aluminum (14,000 kWh for a new plant, over 15,000 kWh for an existing one) but also in the cost of this electricity (Springborn-Townsend: 16 mills/kWh in Australia, 24 mills in the United States; OECD: 20 mills; Woods-Burrows: 15 mills in 1976; and so on).

[d] We refer the reader to the studies listed to evaluate the consumption per ton of aluminum and the unit cost of different materials such as coal, pitch, aluminum fluoride, cryolite, fluorspar, fuel oil no. 6, and so on.

[e] Supervision here is evaluated at 15 percent of direct labor.

[f] Some studies place the figure at 2.5 percent of investment; others differentiate investment on equipment (3 percent) from investment on buildings (2 percent).

[g] Here we find enormous differences among the studies in the amount of investment (OECD: $3,000/tonne of capacity; Springborn-Townsend: $2,133/ tonne for a new plant and $1,000/tonne for an existing one; Woods-Burrows: $1,730/tonne in 1976—all figures in U.S. dollars), in the number of years used to calculate depreciation costs, in the interest rates used for the actualization, and other costs.

[h] We were not able to isolate the "depreciation and interest" cost factors for this study because interest was grouped with profits and taxes. Not only are these data upwardly biased but they also include elements that are not really part of production costs. It should be noted that this study determines the level of profits sought, interest, and taxes by applying a fixed percentage of 19 percent on the investment cost and a rate of 12 percent on working capital.

[i] As opposed to the other studies, the Woods-Burrows study is for the year 1976. We are assuming that the production cost structure remained stable between 1976 and 1980. It should be noted that the plant was new in 1976; we categorized it as an existing plant because of the four years that subsequently elapsed.

mills/kWh over most of the overseas competitors. Some of the smelters near major markets, however, have an offsetting cost penalty in the freight paid on imported alumina or bauxite. Taken together, the effect of transportation costs is minor.

In examining the competitive position of the Canadian aluminum industry, the focus is on electricity costs. According to a recent OECD study, the major individual source of cost variation by location is the price of electricity. OECD's sensitivity analysis provided it with the following likely cost variations by location (when expressed in terms of a change in the cost of electricity, expressed in mills per kilowatt-hour in 1981 U.S. dollars): 45 mills/kWh for the price of electricity, 18 mills/kWh for the cost of capital, and 14 mills/kWh for the cost of alumina.[10] Therefore, we can safely concentrate on electricity costs in assessing Canada's comparative advantage in the production of aluminum.

The Cost of Electricity

The Canadian aluminum industry is distinguished from most other producing countries' by its complete dependence on hydropower to supply electricity to its aluminum plants. This dependence is a considerable asset because hydropower is the lowest cost source of electricity. As can be seen in table 7-4, despite a capital cost almost as high as that of nuclear energy and much higher than those of oil- or coal-based power sources, the higher capital costs of hydropower are more than counterbalanced by its very low operating costs and zero fuel costs. The estimated cost of production for hydroelectric power is 24 mills/kWh, as compared with 39 mills for nuclear power, 48 mills for coal, and about 66 mills for oil.

Of course, Canada is not the only country with great hydropower resources. The 1974 installed capacities of the United States (17.4 percent), Canada (10.6 percent), the Soviet Union (10.3 percent), Japan (6.5 percent), France (5.1 percent), Italy (4.9 percent), and Norway (4.6 percent) accounted for 60 percent of the world total. The installed and potential capacities together have the following distribution: China (14.6 percent), the Soviet Union (11.9 percent), the United States (8.2 percent), Zaïre (5.8 percent), Canada (4.2 percent), Burma (3.3 percent), and India (3.1 percent).[11] Canada is not the only country with ample actual and potential supplies of hydroelectric power; yet it is certainly among the leaders, ranking second in 1974 for installed capacity and fifth for installed and potential capacity together. Moreover, the vast majority of the undeveloped sites worldwide are located in regions that are remote from

Table 7-4. Indicative Estimates of Power-Generating Costs and Other Factors from Different Primary Energy Sources

	Oil				
Factor	Low-sulfur content	High-sulfur content	Coal	Nuclear power	Hydro
Cost[a]					
Capital	10.8	12.9	17.1	24.8	22.1
Operating	2.5	4.2	5.1	4.2	2.0
Fuel	54.6	47.6	26.0	10.0	—
Total	67.9	64.7	48.2	39.0	24.1
Hypotheses					
Capacity (MW)	600	600	600	1,100	600
Interest (discount) rate (percentage)	10	10	10	10	10
Debt amortization (years)	30	30	30	30	50
Investment (dollars/ kilowatt-hour)	577	692	920	1,331	1,249
Duration of construction (years)	3	3	4	6	5

Source: Organisation for Economic Co-operation and Development, *Aluminum Industry: Energy Aspects of Structural Change,* Statistical Annex (Paris, 1983) table 7, p. 112. Reprinted by permission of OECD.

[a] In 1981 U.S. mills per kilowatt-hour.

consumer markets and, in some cases, remote from raw materials markets as well. In this regard, Canada has a privileged position.

As table 7-5 indicates, Canada's installed electricity generating capacity is not evenly divided among the provinces. First, most of the electricity generating capacity of Quebec, British Columbia, Newfoundland, and Manitoba is hydro; the more expensive thermal capacity is concentrated in Ontario and the other Canadian regions. Second, Quebec's hydroelectric capacity alone represents 44 percent of the Canadian capacity, followed by British Columbia (17.5 percent), Ontario (14 percent), Newfoundland (13 percent), and Manitoba (7 percent).

As of December 31, 1983, Hydro-Quebec, the provincial public electric utility, had an installed capacity of 21,301 megawatts (MW) (of which 92 percent was hydro). In 1983 this capacity produced more than 88 billion kilowatt-hours—or more exactly, 88.3 terawatt-hours (tWh). Since 1972 Hydro-Quebec has also had access to the lion's share of production from the Churchill Falls power plant in Labrador, Newfoundland, which has a nominal power capacity of

Table 7-5. Installed Generating Capacity in Canadian Provinces at End of Year 1983: Electric Utilities and Industrial Establishments
(absolute numbers are nameplate ratings in megawatts)

Fuel type	Quebec	British Columbia	Ontario	Newfoundland	Manitoba	Other provinces, Yukon, and Northwest Territories	Canada[a]
Hydro	22,585[b]	8,997[c]	7,131[d]	6,213[d]	3,641[d]	2,707	51,274
	(44.0)	(17.5)	(13.9)	(12.6)	(7.1)	(5.3)	(100)
Thermal	2,068	1,866	19,777	752	502	13,288	38,253
	(5.4)	(4.9)	(51.7)	(2.0)	(1.3)	(34.7)	(100)
Steam	906	1,418	12,898	503	446	11,461	27,632
Nuclear	685	—	6,140	—	—	680	7,505
Internal combustion	714	109	10	79	32	249	593
Gas turbine	363	339	730	170	24	897	2,523
Total installed	24,653	10,863	26,909	6,965	4,143	18,517	89,527
generating capacity	(27.5)	(12.1)	(30.1)	(7.8)	(4.6)	(20.7)	(100)

Note: Numbers in parentheses are percentages.
Source: Statistics Canada, *Electric Power Statistics,* vol. 2, Annual Statistics, Cat. 57-202 (Ottawa, 1983) pp. 8–9.
[a] Includes confidential data, not available by province.
[b] Capacity divided between Hydro-Quebec (20,011 MW) and industrial establishments (mainly Alcan) (2,574 MW).
[c] Of this capacity, 86 percent (or 7,699 MW) belongs to Hydro-British Columbia; the rest (1,419 MW) is controlled by industrial establishments (Alcan has an installed generating capacity of 896 MW).
[d] Capacity mainly controlled by electric utilities.

5,225 MW. By 1985, with the completion of projects under construction, Hydro-Quebec's installed capacity should increase by 3,021 MW, making a total installed capacity of 24,322 MW, not including the capacity available under a long-term (forty-year) contract with Churchill Falls. (This figure also excludes the installed capacity of nonutility producers in Quebec such as Alcan.)

Hydro-Quebec authorities estimate that another 20,000 MW of the province's hydroelectric potential could be developed both economically and relatively quickly between now and the year 2000. Once this potential is developed, the utility will look for other sources of energy to exploit, such as nuclear power and new forms of energy that are now more expensive than hydroelectric power. Obviously, these 20,000 MW will not be developed for the same cost as, for example, the La Grande complex, which was estimated at 23 mills in 1982. Nonetheless, according to provincial authorities, the development of 25 percent of these megawatts

> . . . can be accomplished at a unit cost comparable to that of the La Grande complex: look at the LG-1 station, la Romaine and several small rivers in the James Bay region. A little over half (60%) is made up of the Grande Baleine and Nottaway-Broadback-Rupert complexes; developing these rivers, due to difficulties previously mentioned at the launching of the James Bay project, will be quite expensive, but should nonetheless have an edge over nuclear power. Lastly, the rest comprises several rivers of less importance, whose development costs should be around the same as those of nuclear power.[12]

Such a scenario is depicted in figure 7-2, which illustrates the costs of the potential hydroelectric capacity in Quebec that could be developed after 1985 relative to the costs of the nuclear power. It is clear from the figure that growing demand will encourage Hydro-Quebec to harness rivers farther and farther away from the markets or resort to alternatives such as nuclear power.

In relation to the potential for expansion of hydro capacity, the growth of demand for electricity has been modest. Electricity sales on the Quebec market are mainly determined by population size and by the level of industrial activity. As these variables have increased little in recent years, the province of Quebec has been seeking new customers for its electric power. Despite the signing of contracts with Ontario, New Brunswick, and states in the northeastern United States, however, the growth of supply is likely to continue to outstrip the growth of demand.

The energy surplus (defined as total capability minus regular load) was over 14 tWh in 1983. It will reach 30.4 tWh in 1987,

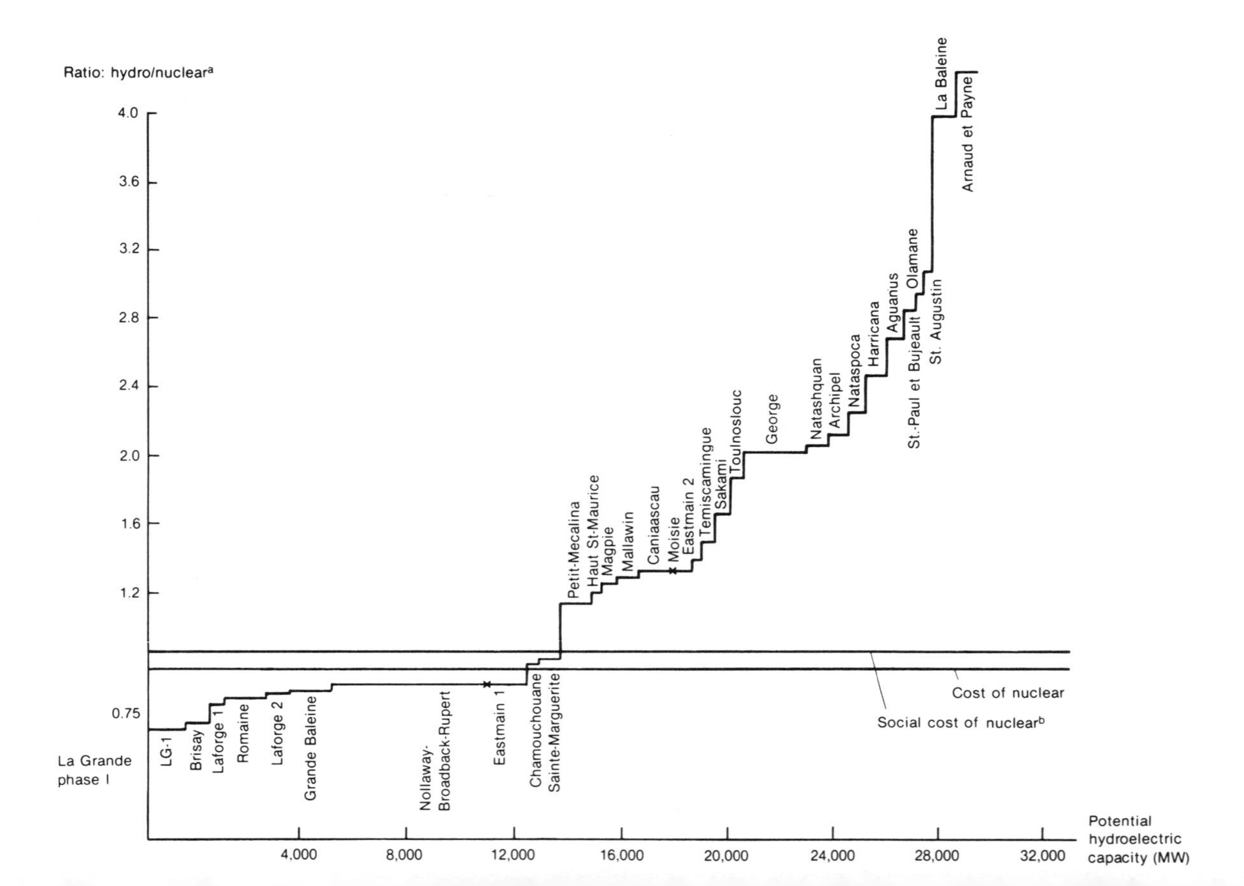
Ratio: hydro/nuclear[a]
4.0
3.6
3.2
2.8
2.4
2.0
1.6
1.2
0.75
La Grande phase I
LG-1
Brisay
Laforge 1
Romaine
Laforge 2
Grande Baleine
Nollaway-Broadback-Rupert
Eastmain 1
Chamouchouane
Sainte-Marguerite
Petit-Mecalina
Haut St-Maurice
Magpie
Mallawin
Caniaascau
Moisie
Eastmain 2
Temiscamingue
Sakami
Toulnoslouc
George
Natashquan
Archipel
Nataspoca
Harricana
Aguanus
St.-Paul et Bujeault
Olamane
St. Augustin
La Baleine
Arnaud et Payne
Cost of nuclear
Social cost of nuclear[b]
4,000
8,000
12,000
16,000
20,000
24,000
28,000
32,000
Potential hydroelectric capacity (MW)

Figure 7-2. Relative cost of Quebec's potential hydroelectric capacity. *Source:* Government of Quebec, *L'électricité: facteur de développement industriel du Québec* (Electricity: A Factor in Industrial Development in Quebec) (Quebec, Ministry of Economic Development, 1980) p. 400.

[a]Unit cost of hydroelectric plant divided by unit cost of a nuclear plant (850 MW CANDU [Canada Duterium Uranium]). The comparison assumes start of operations in 1995. Estimates take into account operating costs of each type of equipment.

[b]Social cost of nuclear energy as estimated by Hydro-Quebec is 10 percent higher than the cost of a nuclear plant.

assuming an average annual growth rate of demand of 3.9 percent, or 52.3 tWh, if we assume an annual rate of 2.9 percent. In the last case, there is enough power to supply a dozen aluminum smelters each with a production capacity of about 250,000 tons. And there will still be, other things being equal, an energy surplus of 30 tWh in 1990 when the new Pechiney-Alumax-SGF smelter and the new capacity of Reynolds are in full operation. The energy excess is estimated to be 19.8 tWh for 1996. The surplus represents firm quantities of energy, which can technically be delivered all year round and over long periods of time; it does not take into account seasonal excess energy, that is, additional energy that could be produced in summer. Thus, the availability of hydroelectric power seems assured in Quebec for at least a period corresponding to the life span of a smelter.[13]

British Columbia's energy situation resembles that of Quebec but at a much lower level. In 1983 B.C. Hydro had a total electric generating capacity of 8,832 MW (of which 85 percent was hydro). This capacity produced in 1983 more than 35.4 tWh of energy, 10 percent of which (3.6 tWh) was exported. B.C. Hydro, like Hydro-Quebec, is faced with surpluses because of overbuilding based on forecasts of demand growth that have proven wildly optimistic. (An average annual growth rate over the next decade of 6.4 percent was predicted in 1981; it was revised downward to 4.8 percent in 1982 and lowered again to 3.9 percent in 1983.)

The B.C. Hydro export surplus has varied in the last ten years: for example, it was zero in 1976, 3.9 tWh in 1978, 0.8 tWh in 1980, and 6.3 tWh in 1982. Exports are possible only when favorable water conditions coincide with export market opportunities. Despite such fluctuations and the fact that the difference between its energy supply and the domestic requirements it serves represents only around 10 percent of the surplus in Quebec, British Columbia may be considered as comparable to Quebec in having hydropower available for the expansion of the aluminum industry. The 1984 completion of the Revelstoke generating plant has added an additional 6.8 tWh of annual energy to B.C. Hydro's system, all of which is surplus to the province's requirements. In the years to come, B.C. Hydro can count on enough energy to supply two or three new aluminum smelters of about 250,000 tons each.

The energy situation in Newfoundland is more complicated and difficult to evaluate. One of the five largest hydroelectric projects in the world is located in this province—Churchill Falls, with a nominal power capacity of 5,225 MW. The ability to use this energy for local needs, exports, or economic development, however, is limited by a contract signed in 1969 between Hydro-Quebec and Newfoundland & Labrador Hydro. Under this agreement,[14] Hydro-Quebec buys 34 tWh annually (around 80 percent of the energy produced) at the rate of just less than 3 mills for a total of about $92 million—less than one-tenth of the power's export value. The purchase price declines to 2.5 mills or a total of $85 million (by the year 2001) and then drops to a flat 2 mills ($68 million) in 2026 for the final twenty-five years of the contract. As owner of the resource, Newfoundland earns a maximum of $6 million annually in rentals and royalties, a figure fixed by a ninety-nine-year water rights agreement that the province is obliged to renew on the same terms and conditions for a second ninety-nine years beginning in 2060. In return for this power, Hydro-Quebec helped finance the $1 billion construction project, shares in the currency exchange risk, and guarantees that the corporation operating the plant will remain solvent for the life of the contract.

In the early 1980s the Newfoundland government made an attempt to take back the water rights for Churchill Falls, but this move has been stopped by a decision of the Supreme Court of Canada confirming the inability of the province to alter the contract. Given the length of time required to resolve such disputes, we can safely suppose that no expansion of the aluminum industry will take place in Newfoundland in the near future.

Manitoba has a total installed generating capacity of 4,140 MW (88 percent of it is hydro), which in 1983 produced more than 21 tWh of energy. Between 1975 and 1983 the energy delivered annually to extraprovincial markets totaled between 2.5 and 7.6 tWh, or between 20 percent and 44 percent of the energy sold by Manitoba Hydro. Such exports are possible because this system is interconnected with Ontario Hydro by two 230,000-volt lines and one 115,000-volt line; with Saskatchewan Power Corporation by three 230,000-volt lines and two 115,000-volt lines; and with utilities in the United States by one 500,000-volt line and two 230,000-volt lines. Manitoba has the surplus power to supply one or two new aluminum smelters. Recently, however, the province announced plans for large-scale sales of power to the United States under long-term contracts lasting up to thirty-five years.[15] These plans will probably rule out this province as a supplier to aluminum smelters unless it constructs additional hydroelectric generating stations in northern Manitoba.

In summary, additional hydroelectric power is and will remain available in Quebec, British Columbia, and probably Manitoba for many years to come. But this availability represents a major advantage only if the power is low in cost relative to alternatives elsewhere. The standard of comparison should be that of marginal opportunity cost in the various countries, a measure that has two quite different meanings: (1) the value to other users of electricity of an incremental unit of electricity consumption and (2) the cost of producing an incremental unit of electricity. When an increase in output is very costly and not a realistic option, the first definition applies. But in Canada an increase in output of electricity is generally feasible, and the second definition thus applies.

The incremental cost—that is, the cost of additional resources to produce more electricity—is illustrated by the load duration curve shown in figure 7-3. This curve is commonly used to represent the total demand for electricity, as well as its variability, over the course of a year. If we suppose that this load duration curve were to remain constant and that the electric utility were to choose a cost-minimizing mix of equipment to satisfy this particular demand, then the marginal cost of electricity would be the cost that must be borne by the electric utility in order permanently to supply one additional kilowatt-hour—that is, if this added kilowatt-hour were to be de-

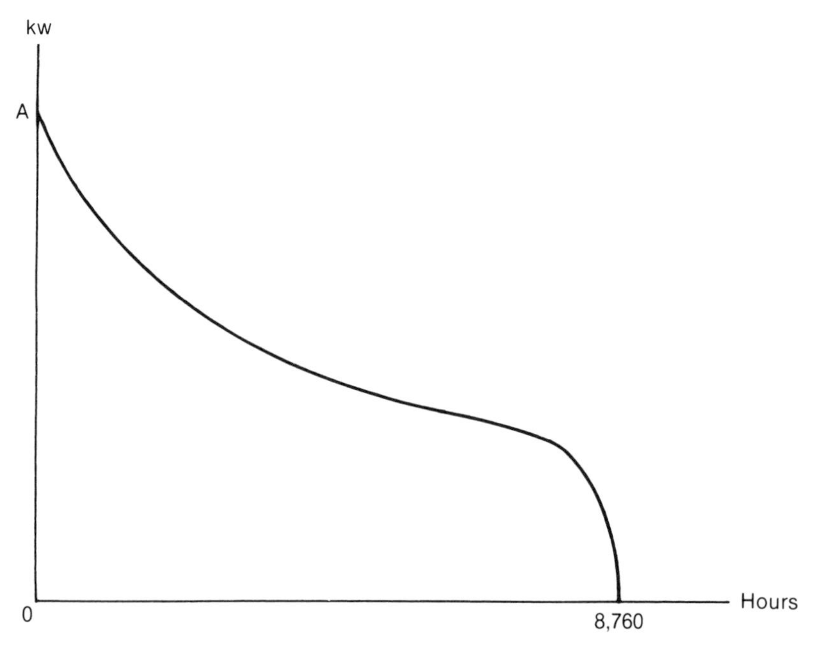

Figure 7-3. Load duration curve.

manded by any user over any one of the 8,760 possible hours in a year. In other words, it is the additional cost of changing the supply by one kilowatt-hour over a base plan, which includes not only the cost of generating this kilowatt-hour but also the cost of its transmission and distribution. It is clear that this incremental cost of providing an additional unit of consumption may differ when the plant cost is fixed (short-run marginal costs) or when adjustments can be made in plant and equipment (long-run marginal costs).[16]

The money value of the marginal costs of a public utility in producing and distributing electricity must be adjusted to take into account the difference between the cost of the utility's factors of production and their value in an alternative use in the economy to obtain the marginal opportunity cost to society. (For example, in a time of continued high unemployment, the amount allowed for labor costs should be lowered to reflect the low value of labor in alternative uses.)

Few studies have been published in Canada that attempt to measure the marginal cost of electricity.[17] Boyer and Martin calculated the cost of electricity in Quebec for the 1976–1990 period following more or less the same line of reasoning as Joskow[18] or Cicchetti, Gillen, and Smolensky.[19] As table 7-6 indicates, Boyer and Martin distinguish three subperiods: pre-James Bay (1976–1979), James Bay (1980–1985), and post-James Bay (1985–1990). These authors attempt to correct the money cost of some Hydro-Quebec factors of production to reflect their opportunity costs. In particular, they take into account the opportunity cost of the corporation's own capital and the funds borrowed in Canada and elsewhere.

Estimates of the marginal cost of electricity have also been provided by the Quebec government's energy management division (see table 7-6). Their marginal costs do not vary over time, as do Boyer and Martin's; instead, they vary according to the different categories of consumers served.[20] These results are not adjusted for the difference in money costs and the value of the factors of production in alternative uses in the economy.

Table 7-6 also presents estimates of the marginal cost of electricity as calculated in a recent study by Bernard and Chatel. Taking into account the limited availability of water power, these authors derive cost-minimizing rules for the choice of generating equipment. These rules form the basis for the estimation of the marginal cost of delivering power to the customer classes (residential, commercial, and industrial) over the peak, intermediate, and base periods. Therefore, the authors recognize that the marginal costs of

Table 7-6. Marginal Opportunity Cost and Marginal Cost of Electricity in Quebec (mills/kWh)

Boyer-Martin Dollars	1976–1979	1980–1985	1985–1990	1976–1990	1976–1985
1976	18.94 (20.83)	25.96 (28.56)	30.03 (33.03)	25.44[a] (27.98)	24.04 (26.44)
1979	19.52 (21.48)	26.77 (29.45)	30.96 (34.06)	26.23 (28.85)	26.77 (29.45)
1981[b]	23.81 (26.19)	32.65 (35.92)	37.76 (41.54)	31.98 (35.18)	30.22 (33.24)

Energy Management Division Type of use	1978	1981	Rate charged (1981)	Rate/marginal cost (1981)
Domestic	24.5	29.2	25.0	0.86
General use—small users	19.3	22.5	34.2	1.52
General use—medium-sized users	16.7	20.0	25.0	1.25
General use—large users	13.1	15.8	13.3	0.84

Bernard-Chatel[c]

Type of use	Marginal cost			Average Price	Rate/marginal cost		
	Peak	Intermediate	Base		Peak	Intermediate	Base
Residential	259.6	18.8	16.8	19.7	0.08	1.05	1.17
Commercial	253.1	18.5	16.6	26.2	0.10	1.42	1.58
Industrial	215.9	16.2	14.9	14.7	0.07	0.91	0.99

Sources: M. Boyer and F. Martin, "Le cout de l'électricité au Québec, 1976–1990" (The Cost of Electricity in Quebec), L'Actualité Economique vol. 54, no. 4 (October–December 1978) pp. 457–561; Government of Quebec, L'électricité-facteur de développement industriel au Québec (Electricity: A Factor in Industrial Development in Quebec) (Quebec, Ministry of Economic Development, 1980) p. 72; and J. T. Bernard and J. Chatel, "The Application of Marginal Cost Pricing Principles to a Hydro-electric System: The Case of Hydro-Québec," *GREEN Research* Papers 8406 (Quebec, Université Laval, March 1984) p. 29.

[a] This cost does not take into account energy losses that can be caused by transporting electricity. If this factor were included, say the authors, the marginal opportunity cost would rise by approximately 10 percent, or from 25.44 to 27.98 mills/kWh (in 1976 dollars), as shown in the figures in parentheses.

[b] Previous line data adjusted by the gross domestic product (GDP) deflator.

[c] 1980 dollars.

supplying electricity to final users depend both on the period when the power is demanded and on the voltage level at which the power is delivered (the latter influences not only the transmission and distribution network but also the associated power losses).

Because institutional and geographical characteristics are to a large extent comparable, estimates of the marginal cost and marginal opportunity cost of electricity in Quebec, as presented in table 7-6, are good approximations of the power costs for other hydro-rich regions such as Manitoba and British Columbia. Table 7-6 brings out the following important features:

- First, even though it is risky to compare the results of three studies that differ in their methods of calculation, the estimates are not too far apart. In general, Boyer and Martin's estimates (which take into account the alternative use in the economy of some of Hydro-Quebec's factors of production)[21] fall in the upper range of the spread in marginal costs calculated by the Quebec energy management division, whereas Bernard and Chatel's estimates tend to fall in the lower range. After comparing the different estimates of table 7-6, we believe that the marginal opportunity cost of electricity in Quebec (and in Canada in general) is in the range of 25–30 mills/kWh for 1980–1985; the cost for large power users is probably between 17–22 mills/kWh. The lower number reflects savings in distribution costs from serving larger consumers.
- Second, the estimates of Boyer-Martin and Bernard-Chatel reflect not the short-run but the long-run marginal cost of electricity. Therefore, they do not take into account the current power surplus of Hydro-Quebec. Short-run marginal costs are difficult to assess under surplus conditions because they depend on the state of the generating capacity, the availability of water over the year, and the transmission capacity, as well as the financial criteria under which Hydro-Quebec is operating.[22] Information with respect to these variables is not readily available.
- Third, the average rate charged to large or industrial users is lower than the corresponding marginal cost of electricity by about 10 percent. This rate is much lower than the marginal opportunity cost estimated by Boyer and Martin. The rates charged by Hydro-Quebec to commercial users (as well as to small or medium-sized users) are generally higher than the utility's marginal cost of electricity. Finally, rates do not differ

substantially from marginal costs for residential users. The negative difference between rate and marginal cost is higher in intermediate and peak periods than in the base period. The government justifies the gap between prices and costs of industrial users by the economic benfits of the industrial users' production activities and by considerations linked to international competition. Although the second argument has some merit in certain cases, the argument of economic benefits does not hold true unless alternative uses for this electricity, such as exporting it, are also evaluated. The question of alternative uses raises the important issue of the opportunity cost of using electricity for aluminum smelting, a question addressed later in a subsequent section on public policy.

- Fourth, little is known about the marginal opportunity costs of electricity in other countries, and the only available point of reference remains the rates charged there. Although its marginal opportunity cost for electricity consumed by large power users is about 17–22 mills/kWh, Hydro-Quebec charges rates of around 15 mills/kWh (1982 dollars). This rate is half that charged by the Bonneville Power Administration in the United States, almost equivalent to the rate mentioned for Brazil (14 mills/kWh)[23] and near the lower end of the price range suggested for Australia (15–19 mills).[24] Moreover, Quebec's advantage is increased because as of 1984 the rate is cut in half in the first five years of a smelter's operation.

Canada, then, has low electricity costs, compared with other potential and actual producing countries, an advantage that does not seem to be offset by higher relative costs for other production factors.[25] Two or three provinces of Canada can supply reliable and low-cost energy supplies for more than half a dozen new smelters for a period corresponding to the average depreciation life of a smelter.

Canada's low electricity prices are the result of three factors: (1) abundant hydroelectric potential, which provides low-cost electricity; (2) an overbuilding of hydroelectric capacity such that the marginal opportunity cost of providing electric power to large users will exclude for a number of years the high capital costs associated with hydro construction; and (3) the willingness of Quebec and some other provinces to provide industry with power at low prices on the presumption that new industry generates external social benefits. It is difficult to assess the relative importance of these three factors. Our qualitative judgment is that even without public subsidy, Quebec would still benefit from low relative energy prices because these reflect the province's long-run low marginal cost of producing electricity (as determined by the first factor). As for the other two factors, we believe that both reasons must be retained in explaining the short-run, low relative energy prices but that their importance will diminish from now until the beginning of the next century.

Comparative Advantage and New Investments

Canada now ranks with Australia and Brazil at the top of the list of nations that enjoy relatively low-cost power. It is hardly surprising, then, that producers recently have shown an interest in increasing their capacity in Canada or that producers not now operating in Canada are interested in building smelters there. Recent investments (Alcan at Grande-Baie, Quebec: 171,000 tons; Pechiney at Bécancour, Quebec: 230,000 tons; Reynolds at Baie-Comeau, Quebec: an expansion of 125,000 tons) and planned investments (Alcan at Vanderhoof, British Columbia: 230,000 tons; Alcan at Laterrière, Quebec: 230,000 tons; Kaiser and Alusuisse at Sept-Iles or Lauzon or Deschambault, Quebec; Alcoa in Manitoba) can be explained by a change in production costs that works to the advantage of Canada as a whole and Quebec in particular. The completion of these investments will double Canada's aluminum production capacity and will affect world trading patterns for bauxite, alumina, and primary aluminum by the start of the 1990s. Therefore, we may conclude that to a large extent the trend in the Canadian smelting capacity during the 1980s does track the shifts in comparative advantages.

As pointed out earlier, Canada's comparative advantage in aluminum production is not a permanent fixture in the nation's history. Although its position was strong early in this century, during the 1920s, after World War II, and now again in the 1980s, the same cannot be said for the 1960s and 1970s. In fact, between 1969–1971 and 1976–1978, Canada's share of world aluminum exports declined in comparison to its share of world exports of manufactured products.[26]

Similarly, there was a decrease in Canada's share of world primary aluminum capacity (from 23 percent to 8.6 percent) from 1960–1979. Several factors account for the decline. First, access to secure, abundant, and low-cost supplies of electricity was less important during that period in determining a region's comparative advantage in the production of primary aluminum. The range of power costs of hydro and oil was much smaller. Also, electricity then was representing between 12 and 15 percent of total production costs and had a much smaller part in explaining international cost variations in the production of the light metal. The "greenfield" or new investment projects tended to be located during that period in countries with a flourishing domestic market for aluminum.

Second, Alcan, the dominant Canadian producer, initiated no greenfield investments during that period because the Kitimat smelter built in 1954 provided substantial excess capacity. Also, Alcan adopted a strategy of increasing the integration of its operations, with the result that investments were directed to the fabricating stage.

Third, there was no federal or provincial public policy aimed at attracting industries to Canada for which the energy costs were a major location factor. The various provincial governments were much more interested in attracting more labor-intensive industries such as automobiles or aircraft. Many other governments, on the other hand, were willing to help the aluminum industry in order to solve balance of payments problems (the United Kingdom) or increase their usage of domestic resources (Australia).

It is difficult to assess whether Canada during the 1960s and 1970s had a comparative advantage in the production of aluminum. The evidence we have suggests that if such a comparative advantage ever existed, it was much lower than in the 1980s.

Given the new aluminum production cost structure and the comparative advantage it gives to Canada, we turn now to the particular policies Canada has adopted to facilitate (or perhaps slow down) the adjustment of producers to these new realities in the international aluminum industry.

GOVERNMENT POLICIES AND THE ALUMINUM INDUSTRY

Canada has no government policies specifically directed at its aluminum industry. Yet the government frequently adopts policies that can influence (either positively or negatively) how aluminum producers adapt to the new structure of the international aluminum industry. These policies touch on areas that include foreign investment control, energy, regional development, labor policy, and environmental protection—at both the federal and provincial levels. In this section, we evaluate governmental measures that can influence the process of adaptation to the changes in comparative advantage for the aluminum industry. We also examine costs and benefits for the national economy and discuss the impact they may have had on the establishment of aluminum capacity in Canada.

The Canadian Foreign Investment Review Agency and the National Energy Program

It has often been argued that the Canadian Foreign Investment Review Agency (FIRA) and the National Energy Program have been detrimental to the Canadian economy in general and its nonferrous metals industry in particular. Although this is not the forum for an evaluation of such statements, no evidence was found of the negative effects of these policies on the Canadian aluminum industry.

FIRA was set up by the federal government in 1973, with the backing of public opinion that favored government intervention in the area of foreign investment in Canada. The main goal of this monitoring agency was not to prevent such investment but rather to ensure that foreign investment generated maximum benefits for the Canadian economy. The bill setting up the agency laid down different criteria by which a proposed investment should be evaluated.[27]

After a decade of operation, it is difficult to avoid the conclusion that FIRA is a paper tiger and is rarely invoked to prevent projects that are submitted to its scrutiny; 88 percent of the foreign investments proposed between April 1974 and March 31, 1977, were approved (excluding proposals withdrawn by investors). The figure rose to 89.3 percent in 1980; it was 87 percent in 1981 and 92 percent in 1982. Even in the mining sector, the agency's directives seem more liberal than those of its Australian counterpart.[28]

The effects of this monitoring agency on the Canadian aluminum industry appear to be almost nonexistent. We have found no cases in which an aluminum producer who wanted to set up shop in Canada, or even to buy national interests, was deterred by problems raised by the agency.

The National Energy Program of October 28, 1980 (which was modified in 1982 in terms of the mechanisms of government intervention but not in terms of the basic principles), was to be a general response to the economic and political problems raised by Canada's energy situation—in particular the questions of energy pricing and the distribution of economic rents. At its core were three basic principles: (1) to ensure Canadians security in supplies and independence from the world oil market; (2) to allow all Canadians an opportunity to participate in the energy sector in general and the oil industry in particular, as well as to share the benefits of industry expansion; and (3) to set up a petroleum-pricing and revenue-sharing regime that recognized the requirement of fairness to all Canadians, wherever they might live.

Again, the energy program has had little impact on the Canadian aluminum industry, in large part because Canadian aluminum producers use relatively little oil or natural gas. There is perhaps an indirect effect in that by preventing Canadian oil prices from catching up with the international price, the program prompts hydroelectric power producers to offer their customers rates that are lower than they would be if oil prices were at the international level. The result is additions to aluminum production capacity and the attraction to this sector of a greater volume of resources than would be justified by the opportunity costs of energy.

It is difficult to evaluate today the effects of the National Energy Program of 1980. By stopping the rise in the international price of oil the recession changed the basic "givens" of the problem, creating the illusion of a permanent overabundance of energy resources. This illusion persists in Canada today despite the indefinite postponement of oil sands megaprojects and technical problems on offshore projects. In this context, reservations about the type of industrial structure that would result from the energy program and the danger of attracting more resources to energy-intensive sectors than would be justified by the long-term equilibrium price seem academic or futile. But it is interesting to note that the same comments were made in the late 1960s when there were concerns about the medium- and long-term effects of oil prices that were lower than their opportunity costs.

The National Energy Program was not designed with the Canadian aluminum industry in mind. Instead, it was meant to achieve oil supply security and Canadian participation in the economic benefits of the oil and gas sector. By the very nature of the problem the policy was a national one as opposed to one for an industry. Its effects on the aluminum industry are at most minor.[29] As we shall see, the same cannot be said of the energy and economic development policies of the provinces, whose effects on the aluminum industry were significant.

Provincial Policies

Quebec

In Quebec, provincial authorities have set out a policy for economic development[30] that clearly defines the wish to use Quebec's comparative advantage in hydro power to attract industries for which electricity is an important location factor. These include the aluminum, magnesium, zinc, newsprint, silicon carbide, and most iron-alloy industries, as well as those that make inorganic chemicals for industrial use.

The economic development policy goes a step further, suggesting that these industries be given special discounts on power rates and guaranteed supplies. Conditions of supply would vary with the level and timetable of proposed investments, their effects (both upstream and downstream) on the Quebec economy, the training and proposed use of local labor, and the participation of Quebec private and public interests. Hydro-Quebec already charges rates lower than marginal costs to consumers in its industrial users category. Although government officials acknowledge the harmful effects of this system of rate setting on the optimum allocation of resources, they stress that the low rates will facilitate the completion of industrial projects in Quebec to a greater degree than if rates more nearly reflected the marginal production costs of electricity. Moreover:

> If these investments generate economic spinoffs whose value is not at least equivalent to, if not higher than, the loss of earnings arising from this less than optimal rate-setting, they would be a poor choice for Quebec society in economic terms. This choice would be even less valid in relation to the export value of electricity, and to some of Quebec's collective goals, particularly in the area of environmental protection, as it would speed up the process of recourse to nuclear energy in the future.[31]

Yet despite these cautionary words, not only is the setting of electricity rates below the marginal opportunity cost for industrial users common practice in Quebec, it is even encouraged. A basic tenet of the current government's economic development policy between 1980 and 1985 has been that economic spinoffs from energy-intensive industries are considered greater than the costs borne by society as a whole. In the case of aluminum, in 1982 the Quebec government offered producers who planned to enlarge their production facilities or invest in new capacities rates of approximately 15 mills/kWh (in 1982 dollars)—less than the long-run marginal cost of electricity sold to large power users, which was estimated to be 17–22 mills/kWh.

In the face of huge estimated electricity surpluses from now until the early 1990s (between 30 and 50 tWh, assuming an annual growth rate for demand of 3 to 4 percent), the Quebec government went even farther and slashed these rates in half for the first five years of operation of new aluminum smelters (if the investments are made in 1984, the rates would be in effect during 1987–1991). They justify such discounts on the grounds that if the surpluses are not sold, the excess hydro capacity will remain idle and water will probably be spilled (they assume a zero opportunity cost for these surpluses). Also, given the high unemployment rate in Quebec, the government wanted to use these discounts as an incentive for producers to accelerate their investment plans.

The discounts were offered despite the fact that the long-run marginal cost of electricity in Quebec compares favorably with that of its major competitors in the world market for the location of new smelting capacity. The government explains the rates by saying that Quebec's competitors do not base their rates on their marginal cost either. Instead, maintains the government, these competitors are giving current and future producers equally attractive discounts to attract new investments. Thus, it is necessary to consolidate and even widen the gap between the production costs of Quebec aluminum companies and those of their competitors.

Of course, these government policies were not designed for the aluminum industry alone. They were meant to be general, with the aim of using the comparative advantage of low electricity costs to attract industries for which the energy factor is a major consideration. This policy was formulated to promote the rapid development of the Quebec economy by benefiting from the turbulent state of the international aluminum industry at a time when an estimated production capacity of 1.5 to 2 million tons became noncompetitive because of the phenomenal increase in energy prices. Presumably, if the same situation had occurred in another industry, such as zinc, newsprint, ferro-alloys, or high technology, the Quebec government would have acted in the same way and offered discounts to gain the anticipated economic spinoffs from those investments as well.

The response to Quebec's policies seems dramatic. Investments announced in 1983 and in the first quarter of 1984 by Reynolds of Canada, Pechiney, and Alcan represented an annual production capacity of approximately 600,000 tons of primary aluminum. The level of investment and its timetable largely reflect the influence of the rates and guaranteed supplies of hydroelectric power. In the case of Alcan, which owns its own hydroelectric dams, the Laterrière project (postponed in May 1985 because of a depressed aluminum market) was doubtless made easier by a long-term agreement on royalties to be paid for using the Peribonka riverbed and water from the Saguenay River.[32]

Despite such an increase in smelting capacity, the possible benefits of new smelters for Quebec should not be exaggerated. If the aluminum producers set up shop in the province for reasons related to the production cost structure (even without discounts, Quebec is a region with a low marginal opportunity cost for electricity), for the same reasons the downstream effects of their investment projects will be quite limited. The production cost structure for aluminum auto parts or aluminum containers and cans is very different from that for ingot production. Transport costs, the absence of a large consumer market, and policies of trading partners (tariffs and other barriers) will be used by aluminum producers to justify fabricating being set up in other regions.

The consequences of this economic development policy adopted at the start of the 1980s by the Quebec government are twofold. First, charging rates below the long-run marginal cost of electricity brings into the province more energy-intensive industries than would have been the case had prices reflected more closely the costs of electricity production. Second, there may be opportunity costs in using electric power for aluminum smelting instead of for other alternatives such as exporting more hydropower to other Canadian provinces or to the industrialized northeastern United States.

Substantial exports of power have, in fact, occurred. In 1983 Hydro-Quebec exported 19.5 tWh (distributed as follows: United States—10.2 tWh; Ontario—5.3 tWh; and New Brunswick—4.0 tWh) or 17.5 percent of the available supply of energy. In 1984 Hydro-Quebec announced a ten-year contract with Vermont for 10 tWh of firm (not surplus) power. Under the contract, Vermont will pay 40 mills/kWh or more than $500 million, a rate below the price of the cheapest alternative for Vermont but above rates for surplus energy of Hydro-Quebec. (The province receives 80 percent of the difference between the cost of production and the cost of the energy it replaces in Vermont.) Also, in June 1984 Hydro-Quebec signed another ten-year contract with the New England Power Pool for 7 tWh annually, or about 5 percent of Hydro-Quebec's generating capacity during 1990–2000. The provincial public electric utility is also negotiating with the New York Power Authority for the sale of firm power. Despite such a marketing effort, however, chances are that surpluses would still be available:

> By 1998, U.S. sales are expected to make up 15% of the total, or about 22 billion kWh (or 22 tWh) a year, vs. the 15 billion kWh going to Ontario and New Brunswick—formerly the biggest export customers. However, this still won't be enough to eat up the estimated 42 billion surplus available from 1987 when James Bay phase one is at full capacity.[33]

Until the mid 1990s, then, Hydro-Quebec will be able to sell all the required quantities of energy to the aluminum producers without preempting exports of firm power to Canadian provinces and northeastern states. In fact, the energy surpluses are such that the discounts (given only for five years and during a period in which it is assumed that the short-run marginal cost of power will be extremely low) seem justified on economic grounds. Although some may interpret such a pricing policy as bad business—because Quebec may be "leaving money on the table," which is unnecessary, given its comparative advantage in the production of aluminum—it may be good economics in the long run if Hydro-Quebec returns to rates reflecting the long-run marginal cost of electricity when the surpluses disappear.

The situation is much more difficult to evaluate for the end of the 1990s and the start of the twenty-first century. Domestic demand may increase at a much higher annual rate than expected, and the same may happen for the northeastern states (due to the closing of some of their nuclear plants and other high-cost energy-producing facilities). In that case, total demand might well outstrip Quebec's energy supply. Such a situation would increase the opportunity cost of power sold to aluminum producers, and the Quebec aluminum industry would be producing more metal than the quantity justified by its current comparative advantage. While this scenario is of course possible, only time can tell if it is an accurate projection of future events.

British Columbia

This western province also has energy surpluses because of the overbuilding of hydro capacity, which was based on forecasts of demand growth that proved to be overly optimistic. For example, the Revelstoke Dam on the Columbia River, which is just now coming into service, provides 1,800 MW of power or 6.8 tWh of annual energy to B.C. Hydro's system; all of it is surplus to the province's needs. In addition, the latest revised forecasts show that British Columbia can expect to be in a power surplus position for at least the next ten years.[34]

Given this surplus situation, what is British Columbia's energy policy? The February 1980 energy policy statement and the May 31, 1984, speech by the minister of energy, mines, and petroleum resources[35] suggest the following answers:

- The B.C. government does not have a policy to build electricity generation projects for the export market. B.C. Hydro has been told by the provincial authorities to postpone plans for new power generation projects. The utility's new role will be to dispose of the surplus and not add to it. Thus, B.C. Hydro must become an active marketing agent.
- Furthermore, the B.C. government does not have an explicit policy of encouraging electricity-intensive industries as part of its industrial development policy. Although B.C. Hydro is offering 30 percent discounts on new industrial sales, this practice is specifically directed to deal with the existing power surplus.
- Since the end of the first quarter of 1984 and for the next six years, B.C. Hydro has been allowed to sell its surplus electricity to the United States as long as the following ground rules are respected: (1) British Columbia's own supply requirements must be satisfied before any power is exported; (2) the export price must be set at least as high as the rate B.C. Hydro applies to incremental sales of electricity to big industrial users in the province (that rate is 18.5 Canadian mills/kWh); and (3) the province will approve firm exports of energy totaling no more than 15 tWh over six years, with an annual maximum of 6 tWh, and interruptible exports of 15 tWh per year minus any firm exports. Since the energy removal certificate was issued, B.C. Hydro has actively pursued new electricity sales, particularly to California.
- Contrary to the province's 1980 energy policy stating that the government will require that all future hydroelectric developments be carried out by B.C. Hydro, the provincial authorities are now encouraging Alcan to bring forward its proposal for a major private hydro development in the north—the Kemano Completion Project.

There are striking differences in British Columbia's and Quebec's reactions to the power surplus situation they have been facing since the end of 1982. First, given its considerable surpluses, the Quebec government has placed no explicit limit on the quantity of firm power its public electric utility may export, whereas such a limit is set for B.C. Hydro. Second, B.C. Hydro's policy always has been and still is to build dams and powerhouses only to meet the needs of British Columbia and not for the export market. The government has also recognized that when the time comes to start building again, it may be wiser to construct smaller projects before contemplating another large-scale hydroelectric development. In the case of Quebec, building for export has not been ruled out, even over the medium term. Phase 2 of the James Bay (La Grande) project may begin as soon as additional sales contracts for firm quantities of energy are signed. Finally, in contrast to British Columbia, Quebec authorities do have an explicit policy of encouraging electricity-intensive industries as part of their industrial development strategy, and they give these industries special discounts on power rates and guaranteed supplies. Conditions of supply vary mainly with the level and timetable of proposed investments and their effects (both upstream and downstream) on the Quebec economy.

In summary, British Columbia authorities favor flexibility in disposing of their power surpluses whereas the Quebec government puts more emphasis on economic development. The former wants to keep all its options open; the latter, on the other hand, is ready to sacrifice a part of this flexibility by giving discounts on electricity rates to convince aluminum producers to accelerate their investment plans.

Other provinces

We have not found any precise economic development policies in other provinces either for energy-intensive industries in general or the aluminum industry in particular. Only Manitoba has the energy to supply one or two new aluminum smelters. But as noted in the previous section, this is perhaps not the only way the Manitoba government intends to use its hydroelectric resources. For example, it has signed agreements-in-principle to sell power to the Northern States Power Company of Minnesota and the Colorado-based Western Area Power Administration in quantities that would necessitate the construction of a second hydroelectric project on the Nelson River.

CONCLUSION

The purpose of this chapter has been to provide some kind of an answer to the following three questions: (1) Does Canada have a comparative advantage in aluminum smelting, and to what extent has it changed in recent years? (2) Do trends in Canada's smelting capacity reflect the change in its comparative advantage? (3) How have public policies helped or hindered the country's industry in adjusting to its new competitive environment?

In answer to the first question, Canada has not always enjoyed a comparative advantage in aluminum production, a case in point being the 1960s and 1970s when Canada's share of Western aluminum production plummeted from 23 percent to 8.5 percent. The increase in the number of government-controlled producers,[36] the lower importance of energy costs in the production cost structure, and high transport costs largely serve to explain this noteworthy dip in Canada's share of the market. The effective protection provided by the tariff structure in major markets for Canadian aluminum and the absence of any Canadian public policy to promote this industry are additional reasons that would explain the shifting of investments in new capacity to other nations.

But the two oil crises changed the production cost structure to make electricity a more significant cost. Endowed with abundant supplies of hydroelectric power and excess generating capacity, Canada in general and Quebec in particular were eager to take advantage of these changes to increase their share of Western aluminum production. Their success in this endeavor was possible because Quebec, British Columbia, and Manitoba could furnish current and future producers with firm, abundant supplies of low-cost electricity sufficient for a dozen new smelters for at least the average depreciation period of an aluminum smelter.

This energy is largely hydroelectric power. There are still rivers that can be harnessed at competitive costs, and the switch to nuclear energy (or hydroelectric power at comparable costs) is not expected to take place until the end of the first decade of the twenty-first century. It is not surprising, then, that these provinces have relatively low energy costs. Their marginal opportunity cost of electricity for large power users has been estimated at about 22 mills/kWh. Although there are no comparable costs available for other producing countries, we believe that Canada is certainly a low-cost location. Electricity in particular has been examined in this regard because electricity is the aluminum production input whose cost varies most among nations.

Thus, our answer to the first question is that Canadian provinces now have a comparative advantage in the production of aluminum because they are endowed with abundant, firm, and generally inexpensive supplies of hydroelectric power.

The answer to the second question is more straightforward. Recent actual and planned investments can be explained by a change in the production cost structure that works to the advantage of Quebec, British Columbia, and Manitoba. The completion of these investments will double Canada's aluminum production capacity. Therefore, we may conclude that the investment in the Canadian smelting capacity during the 1980s does correspond to the changes in Canada's comparative advantages.

Third, Quebec represents a situation in which a producing region with low costs charges electricity rates to industrial users below the long-run marginal opportunity cost of electricity. In the face of huge estimated power surpluses from now until the early 1990s, the Quebec government slashed these rates in half for the first five years of operation of new aluminum smelters. The goals of this explicit industrial development policy are to accelerate new investments (which would have been made later instead) in industries like aluminum in which energy costs are a major factor in location. These actions increased the Canadian comparative advantage and canceled the effects of concessions offered by competitors for new capacity.

Giving short-run discounts on electricity rates can be defended on economic grounds because the surpluses are such that if the power is not sold, the excess hydro capacity will remain unutilized and water will probably be spilled. Until the mid-1990s, Hydro-Quebec will be able to sell all the required quantities of energy to aluminum producers without preempting exports of firm power to other Canadian provinces and northeastern American states. Some have criticized the pricing policy as bad business because Quebec may be offering discounts unnecessarily, given its comparative advantage in the production of aluminum. The long-run effects of charging rates below the marginal opportunity cost of electricity may bring more energy-intensive industries into the province and will probably result in less energy efficiency than when electricity prices reflect more closely the costs of production. Still, given its power surpluses, Quebec's policies are probably good economics, particularly if Hydro-Quebec returns to rates in line with the long-run marginal cost of electricity when the surpluses disappear.

British Columbia does not have an explicit policy of encouraging the siting of electricity-intensive industries in the province. Although B.C. Hydro is offering general discounts (30 percent) on new industrial sales, this practice is specifically directed to deal with the existing surplus power situation. In disposing of these surpluses, provincial authorities want to retain all possible options and do not appear to be ready to sacrifice such flexibility by giving specific discounts to convince aluminum producers to accelerate their investment plans. Given the province's situation and the fact that its power surpluses are not great (they are smaller than Quebec's for example), such a flexible policy may make sense. In the case of the other provinces, we have not found any precise economic development policies for energy-intensive industries in general or the aluminum industry in particular.

The answers to these three questions suggest that Canada dominates as a location for new smelters in the 1980s—a happy ending, but only perhaps. How important such a position will be depends on the rate of growth of the world aluminum industry and the rate at which high-cost smelters elsewhere are closed.

NOTES

1. The reader interested in a more complete treatment of the major stages in the development of the Canadian aluminum industry is referred to I. A. Litvak and C. J. Maule, *Alcan Aluminium Limited: A Case Study* (Ottawa, Ministry of Supply and Services, February 1977); and the Royal Commission on Corporate Concentration, "Case Study No. 13," n.d., chapters 1, 2, and 3.
2. P. Clark, *Rivers of Aluminum: The Story of Alcan,* vol. 1 (Montreal, Aluminum Limited, 1964) pp. 76 and 77.
3. A. L. Dorr and J. E. Tilton, "Bauxite and Aluminum," in C. E. Beigie and A. O. Hero, Jr., eds., *Natural Resources in U.S.–Canada Relations,* vol. 2 (Boulder, Colo., Westview Press, 1980) p. 229.
4. The source of bauxite for this alumina plant has varied over the years. Although British Guiana accounted for 75 percent of Canadian bauxite imports in the mid-1950s, by 1965 the Canadian market was being supplied by Jamaica and British Guiana on a fifty-fifty basis. In the early 1970s, when Guiana nationalized Alcan interests, the company found more bauxite supplies in Guinea (42 percent in 1976). The opening of the alumina plant at Aughinish in Ireland, in which Alcan is a 40

percent shareholder, has again changed the supply structure: the Irish plant will get most of its bauxite supplies from Guinea. The resulting diversion of Guinea bauxite will oblige Canada to diversify its bauxite import sources and add other bauxite-producing countries to its list of suppliers. In 1980 Canada imported 3.35 million tons of bauxite: 51 percent from Brazil, 46 percent from Guinea, and 3 percent from Guiana. The 1981 statistics confirm Brazil's position as Canada's principal bauxite supplier but show Guiana (at 20 percent) and Guinea (at 23 percent) splitting the rest of the market.

5. Besides alumina plants in Quebec and its interest in Queensland Alumina Limited, Alcan has set up alumina plants in Brazil (100 percent interest), India (55.3 percent interest), and Jamaica (93 percent interest). The company also has interests in the San Ciprian plant in Spain (23.5 percent) and two plants in Japan (50 percent each). Its subsidiaries and interests give it guaranteed access to an alumina production capacity of more than 3.75 million tons.
6. Readers interested in the evolution of the cost structure are referred to M. Radetzki, "Long-run Price Prospects for Aluminum and Copper," *Natural Resources Forum* vol. 7, no. 1 (1983) p. 30; and C. Nappi, *Commodity Market Controls: A Historical Review* (Lexington, Mass., Lexington Books, D.C. Heath and Co., 1979) p. 133. The cost of alumina has also risen along with energy prices—18 percent of the total to more than 25 percent; conversely the shares of other raw materials such as coal, pitch, cryolite, and fluorspar have dropped and have gone from 24 percent to less than 10 percent of the total production costs. Labor's share of production costs has decreased slightly; in most cases, it is less than 10 percent. It is more difficult to trace the evolution of capital costs, given the enormous variations in the age of the plants; with older buildings and equipment, depreciation costs are less than with recently built plants. On average, however, capital costs have risen a little more slowly than total production costs, causing a slight drop in the relevant percentage of overall costs in the last decade.
7. It is difficult to find a "typical" cost structure for aluminum production and harder still to discuss how it has evolved over the years. The international aluminum industry comprises a wide range of plants with buildings and equipment of widely differing ages. The plants are supplied by varying sources of energy and alumina and operate in widely differing political and economic climates. Also, it is risky to discuss how the cost structure has developed over a given period of time because we must take into account unpredictable fluctuations in exchange rates, government subsidies that hide the real cost of energy and transport, and the effects of the inflation of the 1970s on the replacement cost of buildings and equipment. Consequently, we should not be surprised to find striking discrepancies in the estimates of total production costs presented in table 7-3.

8. Of course, given the divergent accounting methods used to evaluate depreciation or general overhead, the differences noted in the productivity of electricity and its cost, and the varying availability of raw materials, total production costs can look very different; the average is $1,520 per ton, with a standard deviation of about $200 per ton (all figures in U.S. dollars). To make the estimates easier to compare, we have given not the level of each factor in the cost structure but its percentage of total production costs.
9. Organisation for Economic Co-operation and Development, *Aluminium Industry: Energy Aspects of Structural Change* (Paris, OECD, 1983) p. 49.
10. Ibid., p. 52.
11. Government of Quebec, *L'électricité: facteur de développement industriel au Québec* (Electricity: A Factor in Industrial Development in Quebec) (Quebec, Ministry of Economic Development, 1980) pp. 46–47.
12. Ibid., p. 39.
13. Another way of looking at Hydro-Quebec's privileged position in the energy supply area is to translate its hydroelectric potential into equivalent barrels of oil and compare the figures to known reserves of oil and gas in other Canadian provinces and to the reserves of the ten leading American oil companies. It becomes clear that Hydro-Quebec's estimated hydroelectric potential is equivalent to 57.3 billion barrels (BBL) of oil or 2.4 times the known Canadian reserves of oil and gas, 3.3 times those of Alberta, 23.1 times the Arctic Island reserves, and 38.3 times the British Columbia reserves. In fact, the proven oil and gas reserves of the ten leading American oil companies combined come to only 84.4 percent of Hydro-Quebec's reserves. This estimate of a total economic potential equivalent to 57.3 BBL of oil for Hydro-Quebec applies to the year 1979 and covers previously developed reserves (it does not include for that year the LG-2 dam at James Bay) and reserves that could be developed economically by the year 2000, assuming a lifetime of 100 years for the hydroelectric system. If we add to this base the development of other rivers, a plan that is not currently profitable, Hydro-Quebec's economic potential can be estimated at the equivalent of 80.7 BBL of oil. Interested readers are referred to Kidder, Peabody & Co., "Hydro-Québec," Report of the Corporate Finance Department (New York, November 8, 1983) p. vii; and the National Energy Board, *Reasons for Decision in the Matter of an Application Under the National Energy Board Act of Hydro-Québec* (Ottawa, Ministry of Supply and Services, 1985) appendixes V and VI.
14. For the main details of this contract, see A. Zierler, "Provinces Continue Hydro Dispute," *The Financial Post,* June 2, 1984, p. 22.
15. E. Greenspon, "Power Sale in Works," *The Financial Post,* June 9, 1984, p. 3.
16. It should be noted that most supporters of marginal cost pricing prefer to base power prices on the long-run marginal social cost, particularly

if it does not vary greatly over a five- to eight-year period from the short-run marginal social cost. They point to rate stability and the need to provide consumers with effective long-run signals to help them make intelligent choices on using electric power. This point is particularly well stated by P. L. Joskow, "Marginal Cost Pricing of Electricity," Testimony before the Public Service Commission of New York, case no. 26806 (June 30, 1975) pp. 15–19.

17. M. Boyer and F. Martin, "Le Coût de l'électricité au Québec, 1976–1990" (The Cost of Electricity in Quebec, 1976–1990), *L'Actualité Economique* vol. 54, no. 4 (October–December 1978) pp. 431–462; Government of Quebec, *L'électricité: facteur de développement*, chapter III; and J. T. Bernard and J. Chatel, "The Application of Marginal Cost Pricing Principles to a Hydro-electric System: The Case of Hydro-Québec," GREEN Research Paper 8406 (Quebec, Laval University, May 1984) pp. 1–47.
18. Joskow, "Marginal Cost Pricing," pp. 22–23.
19. C. J. Cicchetti, W. J. Gillen, and P. Smolensky, *The Marginal Cost and Pricing of Electricity: An Applied Approach* (Cambridge, Mass., Ballinger Publishing Company; 1977) pp. 1–34.
20. The differences in marginal costs from one category to another (residential, commercial, industrial) can be explained by the following factors: (1) differing power factor use entails widely differing costs because they do not involve the same capital use (for example, in the industrial category, the factor use reaches 80 percent, whereas the average factor use for the network is just 62 percent; (2) distribution costs are much lower for those who consume more energy due to economies of scale; and (3) rate categories that use the highest voltage energy incur lower costs as a result of voltage reducers and and incidental power losses.
21. Given the enormous reduction in projected investment expenditures announced by Hydro-Quebec in December 1983 (from $47.5 billion projected in 1980 for 1981–1990 to $14.7 billion projected in 1983 for 1984–1993), Boyer and Martin's estimates for the years after 1985 must be viewed with some caution.
22. Bernard and Chatel, "The Application of Marginal Cost Pricing," p. 6.
23. This rate for Brazil represents an annual drop of almost 50 percent relative to 1983. It results from increases in domestic power prices at rates less than the general inflation rate and an undervaluation of Brazilian currency against the U.S. dollar. This illustrates well the hazards of expressing world-wide power prices in one currency: exchange rates, with their current volatility, can and do have a very serious impact in that they may distort the picture and lead to erroneous conclusions.
24. OECD, *Aluminum Industry*, pp. 39–40.

25. There are a number of other factors we have not considered in this examination of Canada's competitive position in the international aluminum industry: for example, the cost of alumina (30 percent of total production costs), capital (34 percent of total costs), labor (7 percent), and transport (4–5 percent). These factors do not vary enough from one country to another to determine the comparative advantages; yet the following comments apply:
   - First, as for alumina, Canada is certainly at a disadvantage in comparison with bauxite-producing countries (we estimate that Canadian producers defray an average weighted price that is about 15 percent higher than for competitors in Brazil and Australia), but they have a slight edge over American producers. This edge is hard to quantify at present, but it could be explained by energy and labor costs.
   - Second, it is harder still to determine the competitive position of aluminum-producing countries in terms of depreciation, interest, and maintenance costs because it is difficult to establish an average age for the plants or to quantify the annual capital cost for each of them. Moreover, when evaluating competitiveness for current output, we have to compare the cost of building new facilities in different countries and not the capital costs of existing facilities because they are sunk costs and by conventional economic theory do not affect the relative position of producers.
   - Finally, when we look at labor and transport costs, we note that the competitive positions of Canadian and American producers are not far apart; the same holds true regarding transport costs for Canadian (and American) producers and those from Brazil and Australia: the strengths of the former in serving the major consumer markets are partially canceled out by additional transport costs for importing bauxite or alumina.
26. United Nations Industrial Development Organization, *Changing Patterns of Trade in World Industry: An Empirical Study on Revealed Comparative Advantage* (New York, United Nations, 1982) pp. 33–203.
27. See Business International Corporation, *Operating in a Changing Canada: A Rich Market, an Uncertain Future* (New York, BIC, 1978) p. 104.
28. Readers who would like to see a further comparison of monitoring mechanisms are referred to the Foreign Investment Review Agency, *A Comparison of Foreign Investment Controls in Canada and Australia,* FIRA Papers no. 5 (Ottawa, Government of Canada, April 1979).
29. A good presentation and evaluation of the National Energy Program can be found in W. Dobson, *Canada's Energy Policy Debate,* Observation no. 23, (Montreal, C. D. Howe Institute, May 1981).
30. Government of Quebec, *Bâtir le Québec: Enoncé de politique économique* (Building the Quebec Nation: A Statement of Economic Policies) (Quebec, Ministry of Economics Development, 1979).

31. Government of Quebec, "L'électricité: facteur de développement," pp. 75–76.
32. The final hurdle for the Laterrière construction start was the issuance of hydraulic leases by the Quebec government as well as pledges to adopt legislation confirming the leases. Alcan has been granted a fifty-year lease for the hydroelectric resources, retroactive to January 1, 1984, providing for two types of payment. The first, representing 90 percent of the rent, is pegged to Hydro-Quebec's basic industrial rate, which is expected to increase by 7 percent per year over the first twenty-five years of the lease. The second form of payment, which is 10 percent of the rent, is indexed to Alcan's price for aluminum and is expected to increase by 6 percent per year over the first twenty-five years. Alcan will pay an estimated total of $662 million to Quebec for use of the hydroelectric resources.
33. S. Yukin, "Growing Demand Could Fuel Hydro's Expansion at James Bay," *The Financial Post,* May 19, 1984, p. 28.
34. The reader is referred to the Ministry of Energy, Mines, and Petroleum Resources, "Energy Removal Certificate: B. C. Hydro Wins Electricity Export Approval," News release (Province of British Columbia, March 21, 1984); and British Columbia Hydro and Power Authority, "1983 Ten Year Electric System Plans and Evaluations (Summary Outline): 1 April, 1983, to 31 March, 1993" (System Engineering Division, December 1982).
35. See the Ministry of Energy, Mines, and Petroleum Resources, *An Energy Policy Statement* (Victoria, Province of British Columbia, February 1980); and the Honourable Stephen Rogers, Notes for a Speech by B.C. Minister of Energy, Mines and Petroleum Resources to the Northwest Public Power Association Annual Meeting (Victoria, May 31, 1984).
36. Government officials encouraged the development of these smelters not only because aluminum was considered a strategic metal but also because the plants favored the use of local resources such as bauxite or alumina and electric power.

INDEX